W9-APS-787

PROCESS REENGINEERING

Also available from ASQC Quality Press

Reengineering the Organization: A Step-by-Step Approach to Corporate Revitalization
Jeffrey N. Lowenthal

Reengineering the Factory: A Primer for World-Class Manufacturing
A. Richard Shores

Business Process Improvement
H. James Harrington

Management by Policy: How Companies Focus Their Total Quality Efforts to Achieve Competitive Advantage
Brendan Collins and Ernest Huge

Untangling Organizational Gridlock: Strategies for Building a Customer Focus
Michele L. Bechtell

To request a complimentary catalog of publications,
call 800-248-1946.

PROCESS REENGINEERING
THE KEY TO ACHIEVING BREAKTHROUGH SUCCESS

Lon Roberts

ASQC Quality Press
Milwaukee, Wisconsin

Process Reengineering: The Key to Achieving Breakthrough Success
Lon Roberts

Library of Congress Cataloging-in-Publication Data

Roberts, Lon.
Process reengineering: the key to achieving breakthrough success / Lon Roberts.
p. cm.
Includes bibliographical references and index.
ISBN 0-87389-274-7 (acid-free paper)
1. Organizational change. 2. Organizational effectiveness. 3. Quality control. 4. Industrial organization. 5. Production planning. I. Title.
HD58.8.R633 1994
658.4'063—dc20 93-45264
CIP

10 9 8 7 6 5 4 3 2

ISBN 0-87389-274-7

Acquisitions Editor: Susan Westergard
Project Editor: Kelley Cardinal
Production Editor: Annette Wall
Marketing Administrator: Mark Olson
Set in Garamond and Univers by Montgomery Media, Inc..
Cover design by Montgomery Media, Inc.
Printed and bound by BookCrafters, Inc.

ASQC Mission: To facilitate continuous improvement and increase customer satisfaction by identifying, communicating, and promoting the use of quality principles, concepts, and technologies; and thereby be recognized throughout the world as the leading authority on, and champion for, quality.

For a free copy of the ASQC Quality Press Publications Catalog, including ASQC membership information, call 800-248-1946.

Printed in the United States of America

Printed on acid-free recycled paper

ASQC
Quality Press
611 East Wisconsin Avenue
Milwaukee, Wisconsin 53202

For all that she is
and all that she is to become,
this book is dedicated to my granddaughter
Morgan Ashleigh.

CONTENTS

FOREWORD

Anyone involved in a management role in a business enterprise should know and understand *process reengineering.* It is a term that is being used frequently in business circles these days. Why? Because the very survival of many firms depends upon how well they manage change to meet customer needs and unprecedented competitive demands. But beyond that, those firms that are alert to opportunities, are quick to respond to customer needs, and can deliver value to their customers will be the clear winners in the future.

Lon Roberts' *Process Reengineering: The Key to Achieving Breakthrough Success* is must reading for those of us having any amount of management responsibility. It is clear that the long-standing, heavy reliance on traditional organizational structure and philosophy no longer applies in today's fast-paced business environment. We are now seeing the emergence of the cross-functional team approach as an effective means of delivering customer value.

This book goes well beyond capturing the essence of process reengineering. In addition to describing what really constitutes process reengineering, Lon Roberts presents the much more difficult "how to" with a simple straightforward style from the beginning of the process through the transition period to completion, with completion defined as the future base for continuous improvement.

PROCESS REENGINEERING

If you are searching for quantum leaps in your organization in the typically difficult area of white-collar productivity, this book will serve as your professional advisor on how to go about the process. As is so often the case, top management must not only embrace the concept of process reengineering but also must be its champion.

Today's customer has never been more discerning or demanding and will only pay for value received. Process reengineering gives us a great assist in assuring that we provide our customers with "cheque book" value.

Bob Caster
President & CEO
Lennox Industries (Canada) Ltd.

PREFACE

Process Reengineering: The Key to Achieving Breakthrough Success examines the process reengineering philosophy and then describes a framework for linking the philosophy to an action plan for implementing a reengineering project. Certain tools and techniques that may be brought to bear on the project are also introduced along the way. Consequently, the book offers a systematic approach to process reengineering rather than a collection of cases and anecdotes. Nevertheless, certain cases are profiled—most notably in chapter 14—to showcase the range of applications and to drive home some important conclusions.

The book also attempts to present a balanced approach to business process reengineering rather than dogmatically insisting on the "decimate and re-create" approach espoused by some. Though the various pundits may all agree on the end result—that is, radical process reform that goes beyond the limits of incremental or evolutionary improvement—this book is grounded in the notion that "starting over with a clean sheet of paper" is often unnecessary, overly risky, and the very reason why many recoil at the thought of reengineering their business processes. At least as far as process reform is concerned, the end should always justify the means and, as with any investment in change, offer a potential payoff commensurate with the risks involved.

Philosophical differences notwithstanding, process reengineering, by definition, does require that business processes be reexamined from a fundamentally new perspective. By so doing, managers may in fact discover that a total redo of the process is justifiable or even essential. This most extreme case should at least be examined as one alternative to achieving radical process reform, whether it proves tenable or not in the end. From this vantage point, trade-offs can be weighed in terms of costs and risks in relation to benefits. Indeed, one hazards squelching the kind of creative thinking that the process reengineering paradigm demands if this basic question is not pursued: How might this particular problem be solved if we were starting anew?

As with any framework, managers must be aware that circumstances specific to the organization will dictate how and if certain aspects apply. For a multitude of reasons, such as fundamental differences in corporate cultures, what works for one organization may generate marginal benefits, or even negative results, in another. This, of course, is the inherent danger of being overly prescriptive and the primary reason why a "cookbook approach" to process reengineering is simply not feasible. Put in other terms, an inflexible process reengineering strategy is no more desirable than an inflexible business process.

While this book is sufficiently informative to stand alone, it is concise by design. Its format will allow the reader—whether executive or line worker—to get to the point without having to wade through reams of text or perhaps getting bogged down in details in order to understand the "whats" and the "whys," if not always the "hows." References are provided to assist those who desire further information on certain topics. Others may find it necessary to secure support from trained specialists to apply certain advanced tools or techniques or, for instance, to implement technology-based reengineering solutions.

Since the book offers a systematic approach to process reengineering, some organizations may wish to use it as a text to support a training program on process reengineering, process improvement, or business process management. Others may find it beneficial in establishing a set of guiding principles and a common language for the reengineering project team or perhaps an executive steering committee.

The author encourages the reader to offer suggestions for future enhancements and revisions. He may be reached at 214-596-2956.

Lon Roberts
Plano, Texas

CHAPTER 1
INTRODUCTION

We hear a lot of discussion these days about change, including the need to change and, especially, the reluctance to change. According to author Paul Dickson, psychologists even have a word for the fear of change: *tropophobia*.[1] It seems that everywhere we turn, the subject of change is being harped on, some going so far as to suggest we need to learn to thrive on chaos, others suggesting, a bit less rhapsodically, that we need to become adept at riding the waves of change. By now there should be little doubt in anyone's mind that change is indeed a critical issue in an age where technology turnover occurs in a matter of months, where customers expect and can demand more for their money, and where the competition is in relentless pursuit.

Unfortunately, the clarion call for change is seldom framed within a context that allows us to roll up our sleeves and take action. Now that an abundance of converts have bought into the message that a change-culture is critical to survival in every sector of the economy, the evangelists of change offer little practical advice about what needs to be done and how to get it done. Likewise, war stories often leave us with the haunting feeling that critical unknowns and dissimilarities between our situation and theirs limit the value of comparisons alone. In essence, what we need are a set of guiding principles that will ensure the emphasis on change is effectively channeled toward improving the way business gets accomplished, whether it be in the public or private sector. To at least some degree, the quality movement has fulfilled this purpose, even with its sibling rivalries and varying shades of interpretation.

The bad news is that even under the best circumstances, it can take years to build a quality culture within an organization. Even then there are few assurances that the quality-related decisions and plans that are made today will achieve the results we desire three to five years from now. As important as the long-term outlook may be, most organizations simply cannot afford to tolerate waste and inefficiency while slowly, and methodically, forging a quality culture. Furthermore, in terms of competitive advantage, we are becoming increasingly aware that quality alone is now less a competitive tool than an absolute necessity. In the words of Arnold Nemirow, CEO of Wausaw Paper Mills, "Quality is your ticket into the stadium. You can't even come into the game unless you have a quality product and process in place. You have to compete on other dimensions today."[2] With these thoughts in mind, where should the change emphasis be directed within the organization?

Should companies strive to produce better products and provide better services? Most assuredly. That is, if they wish to stay ahead of, or possibly even with, the competition. On the other hand, when product quality is measured in terms of three or four defects per million, for example, it is legitimate to ask, How much *value* is being added, from the customer's point of view, by taking Herculean steps to further improve the defect rate? Certainly, in some cases, such as when life and limb are involved, nothing short of perfection is ever acceptable. In other situations, the cost to achieve a minor degree of incremental improvement should be weighed against the accompanying benefits.

Perhaps, then, companies should focus their attention on enhancing worker productivity. This suggestion also has merit. The investment in human capital has long been neglected by most organizations—a condition that is especially descriptive of U.S. organizations. Fortunately, many of the better-managed companies have become painfully aware of this situation and have started investing an amount equal to 2 percent, or more, of their annual salary budget in human resources development. Nevertheless, when employees are working 12-hour days and 6-day work weeks to try to cover their own job responsibilities as well as those of displaced associates, productivity enhancement alone is unlikely to yield a sufficient return on investment to hold the competition at bay. Nor is this the time to offer trite advice or resort to slogans such as "work smarter, not harder," at least not without running the risk of causing an outright rebellion.

As important as product improvement and productivity enhancement are to a company's competitive position, an additional area of concern deserves equal, sometimes more, consideration: the effectiveness and efficiency of the business processes that support the development and delivery of the organization's products and services. Inefficiencies in business processes abound. They may include such familiar actions as repeatedly cycling an approval form between departments, creating unnecessary red tape, or generating forms that get put into files that nobody ever refers to—simply to name a few. As most everyone is well aware, business processes have a way of creating a life of their own. Furthermore, they have a way of becoming institutionalized and even canonized. Perhaps it's not surprising, then, when companies discover what some have discovered through the school of hard knocks—that fundamental changes in business processes can result in cost reductions and improvements in cycle time that are several orders of magnitude better than that achievable through other means.

In his seminal article in the July-August 1990 edition of the *Harvard Business Review*, Michael Hammer laid much of the groundwork for an approach to—or more appropriately a philosophy and a framework for—radically overhauling and improving business processes.[3] Hammer compared the conventional add-on approach to process improvement to the act of paving cow paths. He also applied the term *reengineering* to the more radical approach, stating, "Reengineering strives to break away from the old rules about how we organize and conduct business . . . . (by) recognizing and rejecting some of them and then finding imaginative new ways to accomplish work."

Others have added qualifiers to the term reengineering to emphasize certain distinctions in meaning and application. For instance, the expression *process reengineering* makes it clear that the focus is on processes, as opposed to products. *Business process reengineering* is the expression some prefer to stress that business processes are the central issue in contrast to production, or manufacturing, processes. At least one organization, Texas Instruments, uses the expression business process engineering to designate " a start-to-finish holistic view of the related set of activities that deliver value to a customer."[4]

This book has adopted the expression *process reengineering* in lieu of *business process reengineering*—although the intended meaning is the same—to avoid any possible misperception that the concept applies only to profit-oriented businesses. Hospitals and certain governmental entities, for instance, are also realizing benefits from process reengineering. The rules of convention also suggest that *re-engineering* (with a hyphen) is preferred over reengineering (sans the hyphen). While the present author has chosen to adhere to the form of the term as used by its originator, those involved in research, where a hyphen can make a considerable difference when performing an on-line word search, should be aware that both forms of spelling are prevalent.

But why the emphasis on seeking to reengineer business processes as opposed to production processes? For these simple reasons.

- Improvements in business processes have not kept up with improvements in production processes over the years. In other words, the margin for improvement is greater.

- Waste and inefficiency are more difficult to detect in business processes than in production processes.

- Business processes typically cut across the functional lines of the organization, giving rise to crossfunctional problems that less commonly affect production processes.

- Business processes often devote as little as 5 percent or less of the available process time to activities that deliver value to the customer.[5]

- Customers are five times more likely to take their business elsewhere because of poor business processes as opposed to poor products.[6]

In total, these factors easily justify the need to consider a radical overhaul of business processes rather than a simple patching up of existing processes or a search for incremental improvements.

As described in subsequent chapters, process reengineering embodies a philosophy as well as a framework for radically overhauling and improving business processes. Since business processes exist in every organization, and process inefficiency knows no bounds, process reengineering is not limited to a particular type or size of organization. Processes that are steeped in organizational bureaucracy represent some of the greatest challenges, and often the greatest opportunities, for putting process reengineering to work.

So where does process reengineering fit in relation to continuous process improvement? In essence, they are complementary rather than opposing approaches to improving business processes. In certain situations, a process might first be reengineered to ensure that it is designed for optimum performance and then incrementally improved over time as the need for fine tuning becomes evident. Certain critical processes may have to be reengineered a number of times throughout their life cycle, since the external environment is continually changing and since processes have a way of taking on a life of their own that, in some cases, may very well work counter to the best interests of the organization. Work teams charged with continuous improvement typically lack the authority, perspective, and/or capability to implement radical changes that can impact other processes, cut across functional lines, or otherwise impact the organization at large. Process reengineering, on the other hand, makes such changes possible.

PROCESS REENGINEERING

Perhaps the following example will help clarify what makes process reengineering fundamentally different from other process improvement strategies. It should also shed some advance light on the rationale behind the tenets presented in chapter 2.

REENGINEERING IN ACTION

This case examines a reengineering project involving a large insurance company and one of its claims-handling processes. The identity of the company has been masked so that the case can be dramatized for instructional purposes.

The company, Omni Life & Indemnity Company (a pseudonym), is a large multiservice provider with 28,000 employees. Sensing competitive pressure, coupled with the need to be more responsive to its customers, Omni decided to overhaul its process for handling claims related to the replacement of automobile glass. The CEO thought she would be able to use any success acquired from this relatively low-risk endeavor as a springboard for undertaking even more ambitious reengineering projects later. Furthermore, the project would give the company an important head start on the learning curve.

The CEO immediately appointed an executive sponsor to shepherd the project. Following some preliminary analysis of the potential payoff and impact, the CEO and the executive sponsor developed a charter for a reengineering project team and collaborated to handpick its members.

Early in the process, the reengineering team created a flowchart of the existing claims-handling process to better understand things as they were. This flowchart is depicted in simplified form in Figure 1.1.

Figure 1.1 indicates the following sequence of events for processing claims under the initial configuration.

1. The client notifies a local independent agent that he or she wishes to file a claim for damaged glass. The client is given a claim form and told to obtain a replacement estimate from a local glass vendor.

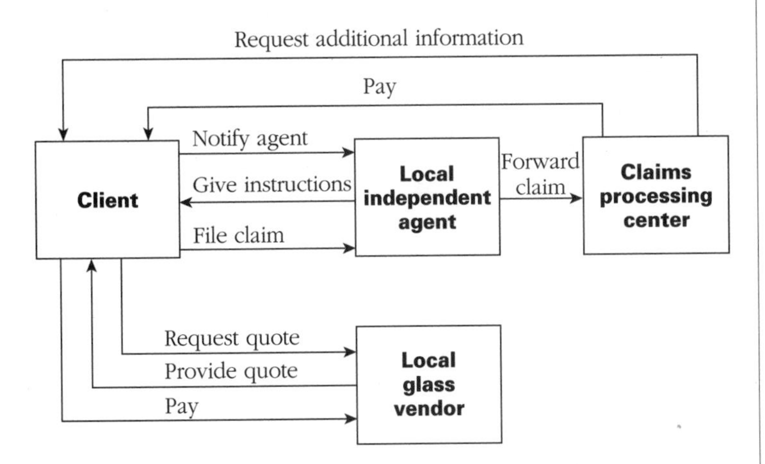

Figure 1.1 Claims processing—initial configuration.

2. Once the client has obtained the estimate and completed the claim form, the independent agent verifies the accuracy of the information and then forwards the claim to one of Omni's regional processing centers.

3. The processing center receives the claim and logs its date and time of arrival. The contents of the claim are then entered into a computer by a data entry clerk (mainly for archiving purposes), after which the form is placed in a hard-copy file and routed, along with the estimate, to a claims representative.

4. If the representative is satisfied with the claim, it is passed along to several others in the processing chain, and a check is eventually issued to the client. But, if there are any problems with the claim, the representative attaches a form letter and mails it back to the client for the necessary corrections.

5. Upon receiving the check, the client can then go to the local glass vendor and have the glass replaced.

The client might have to wait one to two weeks before being able to replace his or her automobile glass under the initial configuration. If the glass was broken on a weekend, the process could take several days longer.

Given the charter to come up with a radical overhaul of the process, the reengineering project team recommended the solution shown in Figure 1.2. The team accomplished this after evaluating a number of process configurations in terms of the costs versus benefits resulting from each configuration.

Structural as well as procedural changes were put in place. Some of these, especially the procedural changes, are not entirely evident simply by contrasting the flowcharts.

The procedural changes involved the following:

- The claims representative was given final authority to approve a claim.

- A long-term relationship was established with a select number of glass vendors, enabling the company to leverage its purchasing power and to pay the vendor directly. Furthermore, since prices are now prenegotiated, the need to obtain an estimate from the vendor no longer exists.

- Rather than going through a local agent, obtaining an estimate, and filling out a form, the client now simply contacts the processing center directly by phone to register a claim.

The structural changes are manifest in the following sequence of events, which describe the flow of the process.

1. Using a newly installed 24-hour hot line, the client speaks directly with a claims representative at one of Omni's regional processing centers.

2. The claims representative gathers the pertinent information over the phone, enters the data into the

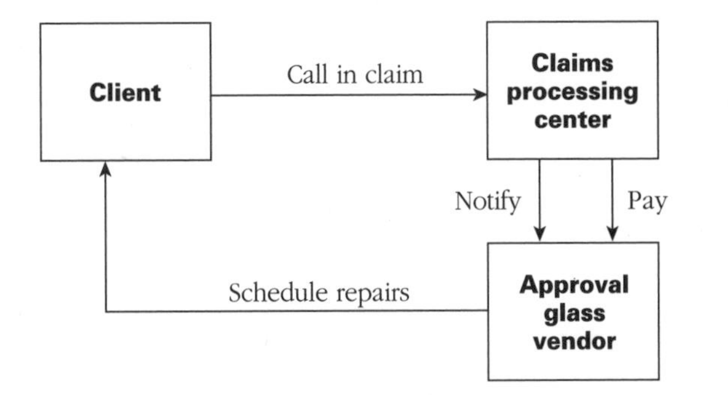

Figure 1.2 Claims processing—reengineered configuration.

computer, and resolves any problems relating to the claim on the spot. The representative then tells the client to expect a call from a certain glass vendor, who will make arrangements to repair the glass at the client's convenience.

3. Since the claim now exists as an electronic file that can be shared through a local area network (LAN), the accounting department can immediately begin processing a check that will be sent directly to the local glass vendor.

A number of significant benefits—some more quantifiable than others—resulted from reengineering this process.

- The client can now have his or her glass repaired in as little as 24 hours versus 10 days. This represents a 90 percent improvement in the process cycle time.
- The client, now has less work to do since a single phone call sets the process in motion. Also, the client is no longer required to obtain a repair estimate.
- Problems are handled at the time the call is initiated, thus preventing further delays in the process.

- The problem of lost or mishandled claims has virtually disappeared.

- The claim now passes through fewer hands, resulting in lower costs because fewer people are involved in the processing chain.

- By establishing a long-term relationship with a select number of glass vendors, the company has been able to leverage its purchasing power to obtain a 30 percent to 40 percent savings on paid claims.

- Since fewer glass vendors are involved, a consolidated monthly payment can be made to each approved vendor, resulting in additional savings in handling costs. Omni now issues 98 percent fewer checks.

- By dealing with preapproved glass vendors, the company can be assured of consistent and reliable service.

- The claims representatives feel a greater sense of ownership in the process, since they now have broader responsibilities and expanded approval authority.

Note that the reengineering team framed the situation in these terms: "How do we settle claims in a manner that will cause the least impact on the client while minimizing the cost to do so?" If instead they had asked, "How can we streamline the existing process to make it more efficient?" their inquiry might not have removed the independent agent from the processing chain. Nor is it likely that any of the other efficiency improvements involving the elimination of jobs could have emanated from a process improvement team consisting solely of members having a vested interest in the outcome.

Could further efficiency improvements still be realized in this process? Most assuredly! But this is where the continuous process improvement mechanism takes over.

COURAGE TO REFORM

Change is never easy. The organization that finds change difficult to take in small doses may find it unpalatable to even imagine the degree of change that is characteristic of process reengineering. Nevertheless, virtually every core business process is, at some time, a candidate for radical reform. As difficult as change may be, in light of today's competitive business environment, most organizations do not have the option of choosing whether or not to reform. Reform can be triggered by vision or by desperation. Given any say in the matter, it is almost always better to act on vision rather than react out of desperation. This is where process reengineering works best—as a proactive instrument for change.

REFERENCES

1. Paul Dickson, *Words: A Connoisseur's Collection of Old and New, Weird and Wonderful, Useful and Outlandish Words* (New York: Dell Publishing Co., 1982), p. 227.

2. John A. Byrne, "Paradigms for Postmodern Managers," *Business Week*, 1992 Reinventing America issue (23 Oct. 1992): 62.

3. Michael Hammer, "Reengineering Work: Don't Automate, Obliterate," *Harvard Business Review* 68 (July-Aug. 1990): 104.

4. Texas Instruments, *Business Process Engineering Concepts* (Plano, Tex.: Texas Instruments, 1992).

5. Joseph D. Blackburn, "Time-Based Competition: White-Collar Activities," *Business Horizons 35* (July-Aug. 1992): 97.

6. H. J. Harrington, *Business Process Improvement: The Breakthrough Strategy for Total Quality, Productivity, and Competitiveness* (New York: McGraw-Hill, 1991).

CHAPTER 2
PROCESS REENGINEERING PHILOSOPHY

Several years ago *Reader's Digest* ran an interesting story about a woman who, before baking a ham, always trimmed a small amount off each end of the ham. When her young daughter inquired one day as to why she did this, the woman, thinking for a moment, stated that she was wasn't certain why, but that she had learned the technique by watching her own mother. She thought it had something to do with making the ham cook more evenly throughout, but she would need to verify this with her mother. When the woman later posed the question to her mother, she was surprised to learn that her mother was not certain either why this was done, but that she likewise had learned the technique by watching her mother, the young girl's great-grandmother. When the occasion arose at a family gathering to ask this question of the great-grandmother, she replied, "The only pan I had available was too small for an entire ham...I always had to trim both ends of the ham to make it fit the pan."

This anecdote cuts to the very heart of the issue surrounding process reengineering. The parallels to the way that business processes evolve are recognizable to anyone who has ever worked in an established organization. As in the anecdote, most of us can plod along day after day doing whatever it is we do and seldom even question why. Never mind the fact that (1) there may be a better way of getting the job done or (2) the original reason for performing the task has long since vanished.

But the influence of tradition is only one of a number of reasons that business processes eventually become inefficient. (Others will be examined later.) Furthermore, some processes are inherently inefficient from the beginning. Either way, business suffers to some degree as a result of wasted resources, unnecessary delays, and most damaging of all, unhappy customers. If the situation is sufficiently drastic and a core business process is involved, a process reengineering initiative may well be justified.

DISTINGUISHING CHARACTERISTICS OF A BUSINESS PROCESS

Before the philosophical underpinnings of process reengineering are examined, we need to be clear on what we mean by a *business process.* In turn, we must also define certain related terms and distinguish between a process and a project, since projects represent still another way in which tasks get accomplished within the organization.

A *process* consists of an activity, or a set of interrelated activities, intended to transform one or more inputs—at least some of which represent customer requirements—into one or more outputs that represent solutions from the internal or external customer's point of view. The more complex business processes are likely to cut across the functional or departmental lines of the organization. In fact, points at which the functional lines within a process interface often present opportunities for dramatic improvement in the process.

Organizations typically have numerous business processes in place. Here is a random list of only a few of the more common ones.

- Order processing
- Billing
- Purchasing
- Shipping
- Receiving
- Inventory management
- Auditing
- Business planning
- Warranty and claims processing
- Budget planning
- Accounts payable
- Accounts receivable
- Performance appraisal process
- Proposal development
- Client acquisition
- Credit approval process
- Management development
- Facility-change request process
- Configuration management
- Contracts administration
- Continuous improvement process
- Human resources planning
- New employee processing
- Proficiency certification

Again, this is only a partial list of generic processes we might expect to find in an organization. Other processes exist that are specific to an organization or a certain type of industry or professional specialty. Processes are so prevalent that most of us give little thought to the fact that much of what we do on a daily basis supports a process to some degree or another—at least, that is, until something goes wrong.

The process itself is commonly composed of two or more subprocesses, sometimes referred to as functions, which should not be confused with the functional departments of the organization, since subprocesses often stretch across departmental lines. These subprocesses, or functions, can be further decomposed into tasks, which are executed by people and/or machines. Furthermore, the various processes that work together to support the overall mission of the organization can, in the grand scheme of things, be thought of as subprocesses themselves. (Figure 2.1 illustrates this hierarchy by showing a partial set of processes, subprocesses, and tasks.) These distinctions are especially important because it is possible to optimize the performance of one subprocess while suboptimizing the overall process. The same holds true for tasks in relation to subprocesses and, at the other end of the spectrum, for one process in relation to another. In other words, to some degree or another, any change that is made within a process will have a ripple effect throughout the entire process, or perhaps the entire organization.

Notice from Figure 2.1 that at the subprocess and task levels, an action verb is used in conjunction with a noun to describe the work that is to be accomplished by each subprocess or task. (Here, what is referred to as the subprocess level maps out one entire process; the task level depicts a set of tasks that constitute a single subprocess.) As this example shows, the distinction between a task and a subprocess is not necessarily evident upon simple inspection of the block diagram labels. The distinction is this: typically, a *task* is assignable to an individual, while a *subprocess* consists of a set of tasks that involve more than one person and, as noted earlier, can even cut across the functional lines of the organization.

This difference highlights the fact that when a business is defined in terms of a set of processes that transform customer needs into value-added solutions, the functional departments of the organization often lose much of their identity. In some cases, to the dismay of many, the functional departments even lose their basis for existence—at least, that is, as traditional, semi-isolated entities.

In terms of analyzing process performance, it is important to recognize that the primary issue at the process level relates to *what is accomplished.* At the subprocess level, the emphasis is on *how the process is organized.* And, at the task level, the central issue concerns *how the job gets done.* From a process reengineering perspective, these issues focus the analysis activities as they correspond to each of the three levels.

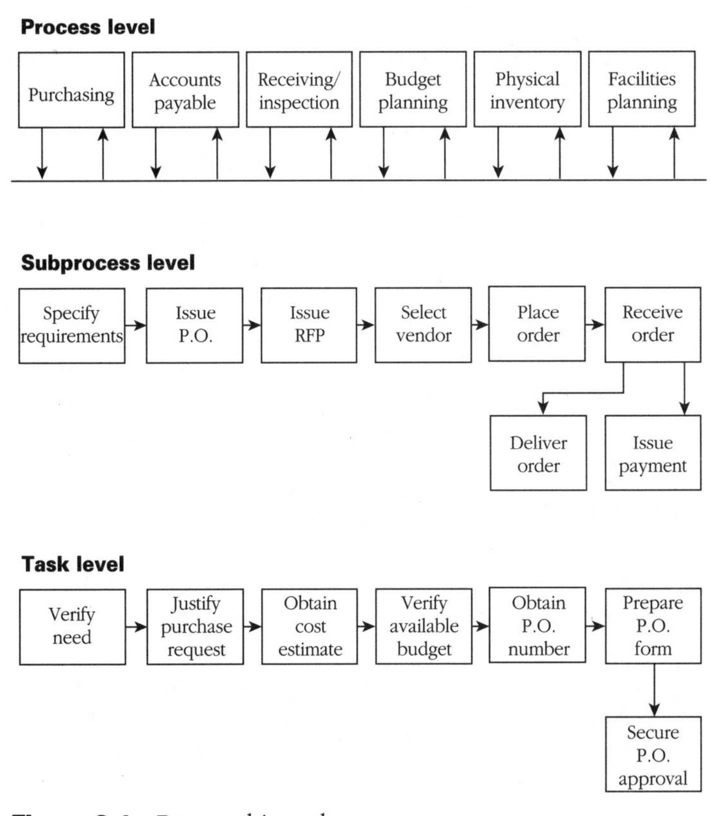

Figure 2.1 Process hierarchy.

Returning to Figure 2.1, notice that the logical flow sequence is depicted by the arrows linking the components in both the subprocess and task level representations. These links typically, but not necessarily, represent some form of data that triggers the need for action from one subprocess to the next or, as the case may be, from one task to the next.

As with all business processes, the processes listed earlier have one thing in common: they all involve the flow and exchange of information in some form or other, although, depending on the process, more tangible items may also pass from one subprocess to the next. Since information handling is such a prevalent aspect of every business process, process reengineering solutions often result in some improvement in the information-sharing links within the process—though this is certainly not always the case, nor is it necessarily the initial driving force for reforming the process. For instance, in the case discussed in chapter 1 involving the Omni Life & Indemnity Company, information shared via the local area network was critical to the efficiency of the work accomplished within the claims processing center. But, if the reengineering team had started with the assumption that the basic problem was one of efficiently sharing information resources, it may not have, for example, decided to leverage the company's purchasing power by working exclusively with a set of preferred vendors.

Note also that while the more formal processes are typically guided by a set of written policies and procedures, this is often not the case for small-scale processes. Ironically, the former—that is, the formal processes—are often the prime candidates for being reengineered. There are at least a couple of reasons why this often holds true.

1. Formal processes typically involve multiple departments and a relatively large number of employees. As a result of their scope and complexity, such processes, usually have the most to offer in terms of net benefits realized from improvement.

2. Because formal processes operate under a set of rigid policies and procedures, they are more likely to be bound to assumptions and/or realities that are no longer valid. While informal processes can also become slaves to tradition, such processes are more likely to be improved as the need for improvement becomes apparent—especially if the organization supports change and does not discourage risk taking.

As mentioned earlier, organizations respond to customer needs by means of processes and/or projects. The distinction between the two is significant. In essence, *processes* are cyclical in that they repeat a set of prescribed tasks on a recurring basis. *Projects,* on the other hand, are more episodic. They typically have a distinct beginning and end. In most cases, no two projects are identical either in terms of the desired results or the way they are designed to accomplish those results. As a further distinction, projects are guided by plans, whereas, processes are guided by procedures, whether they be tacit or explicit. The internal support departments within an organization often function on a project basis if their primary responsibility is one of providing ad hoc support to the various business processes.

As its name implies, *process reengineering* is primarily concerned with improving processes, as opposed to projects. Be aware, however, that process reengineering opportunities are typically planned and administered as projects since (1) they have a specific beginning and end, and (2) no two process reengineering situations are identical. Therefore, those who will be involved in a process reengineering project must understand certain project management concepts as they apply in this context. Keep in mind, however, that process reengineering projects bring an extra dimension of complexity to the project environment since they require greater attention to the dynamics and management of change within the organization. The social and psychological implications of process reengineering are discussed in subsequent chapters.

FUNDAMENTAL TENETS OF PROCESS REENGINEERING

Regardless of how it is ultimately accomplished, process reengineering is always concerned with improving the efficiency and the effectiveness of an overall process. In this context, the following definitions apply.

- *Efficiency* refers to the degree of economy with which the process consumes resources—especially time and money.
- *Effectiveness* is concerned with how well the process actually accomplishes its intended purpose, here again from the customer's point of view.

Certainly, much of what constitutes the concept of process reengineering is not entirely new. Documents describing *work simplification*, for example, can be traced back 40 years and beyond.[1] So why all the current interest in process reengineering? For one, the stark reality is that, with the advent of a global economy, domestic markets have virtually disappeared. From a customer's point of view, sharing a common geographic identity with a product or service provider is insufficient reason to sacrifice quality and value when sources elsewhere are willing and able to do better. As many "solid" U.S. companies can attest from first-hand experience, the gap between what is and what should be will quickly widen beyond control if a business-as-usual attitude prevails. This reality leads us to the first tenet of process reengineering.

Tenet 1: *The customer, and the customer alone, is responsible for defining what constitutes product or service value.* The real winners go so far as saying that not only must customer needs be met, but that customers should be *delighted* with the results.

Related to the pressure from competitive forces is the fact that diminishing profit margins demand that we take a fresh look at the structure of the organization from the top down. There simply is no room for sacred cows and individual interests when the very survival of the organization is at stake. In short, any aspect of the organization that does not add value to the products and/or services demanded by the customer is subject to scrutiny. The same applies to those all-too-common aspects of the organization that not only do not add value, but actually stand in the way of progress—including, perhaps, the organizational structure itself. With this in mind, we can articulate the second tenet.

Tenet 2: *The organization should be structured from the top down to support its value-added processes.* Evidence is mounting that the horizontal process management approach to running the modern business is quickly displacing the vertical command-and-control style of management that has its roots in the turn-of-the-century industrial revolution.

The process reengineering revolution is also being fueled by two other important realities: (1) that improvements in business processes lag far behind improvements that have been made in manufacturing processes and (2) that conventional measures of business success do not adequately track the performance of a true customer-focused organization. Customer satisfaction ratings and process cycle times, for instance, are displacing such indexes as market share and return-on-equity. Even at the highest levels, the measures for success are shifting away from the traditional "bean-counting" indexes of financial management to those that correlate to total customer satisfaction. This shift has the added effect of placing a premium on managers who truly understand the business and can relate to—rather than be detached from—the customer. Furthermore, since process streamlining is the key to becoming more competitive, it also suggests the possibility that cost accounting methods may need to be adopted that reward, rather than punish, process efficiency. These factors lead us to the following tenets.

Tenet 3: *Business processes—the domain of the so-called white-collar worker—hold the potential for quantum leaps in improvement.* Due to their complexity and the difficulty in evaluating them, business processes pose a greater challenge to reform than their manufacturing counterparts. This perhaps explains why improvements in business processes have been virtually neglected until now.

Tenet 4: *Dramatic improvements in cycle times, process costs, and/or customer satisfaction ratings are key indexes of the success of most process reengineering projects.* While the reengineering project may be directed toward only one of these indexes, others may be affected as well. For instance, the reduction in cycle time for a certain process will lead to a more efficient process, which, in turn, translates into lower costs to the customer and consequently a more satisfied customer.

Tenet 5: *The people who directly support the business process should be given a central role in analyzing and redesigning the process.* This is critical for two reasons: (1) those who own and control the process will likely have the best ideas for improving the process, and (2) involvement is a key strategy for overcoming the resistance to change. On the other hand, be aware that an outside change agent may also be needed on the process reengineering team to ensure objectivity.

By now it should be clear that process reengineering involves change on a grand scale. At the very least, a process reengineering project will likely point to the need to revamp a major business process that cuts across the functional lines of the organization. Thus, it must be understood and clearly communicated that (1) the global interests of the organization supersede anything less and that no single process, or subprocess, operates in isolation; and (2) the likelihood of a stalemate is quite possible if turf issues between functional departments cannot be resolved. These concerns lead us to the following tenets.

Tenet 6: *Senior management must be involved throughout the process reengineering project.* There is no middle ground in this respect. Management involvement makes process team members keenly aware that the reengineering effort is directly linked to organizational strategy plans. It also acknowledges that turf issues will arise that may have to be arbitrated by a higher authority.

Tenet 7: *Process reengineering seeks to optimize the performance of the process in relation to other considerations.* Optimization concerns must also be considered at the subprocess level and task level. This will likely include an analysis of the redistributed workload resulting from the reengineered process.

Tenet 8: *Communication and trust are pivotal to the success of the process reengineering project.* Fears and suspicions naturally run high when major changes are being planned. If the people directly involved with the process feel that there is a hidden agenda or that they are being cut out of the communications loop, the project will be doomed from the start.

Process reengineering compares the *as-is* condition of a process with the *as-should-be* condition and then seeks—through redesign, simplification, optimization, and then, if necessary, automation—to improve the process radically in order to bring the as-is closer to the as-should-be. This approach to reengineering a process leads to the following tenets.

Tenet 9: *A carefully planned system of measurements is necessary to establish how well a process is performing and to compare before and after results.* The process team, as well as upper management, should be involved in selecting the parameters to measure and deciding how to benchmark the performance of the process against internal and/or external standards. Care must also be taken to ensure that the various indexes of process performance are internally consistent (that is, not contradictory) throughout. For instance, it makes little sense to evaluate a customer service representative (CSR) on the basis of number of calls taken while avoiding difficult problems that might impact the CSR's performance as measured by this index.

Tenet 10: *Process reengineering begins by asking whether an existing process is essential to the organization. If so, improvement is sought by streamlining the process and then optimizing it in relation to other processes.* Note that while process reengineering may result in a technological solution to a problem, this should not be an initial objective. Form should follow function.

The greatest challenge to successfully reengineering any process is certain to occur during the implementation phase. In terms of impact on the process and people's jobs, little is actually gained or lost until that time. Even a flawlessly designed process will not succeed in the end if those who have a vested interest in the process or its outcome resist change. Therefore, users must be involved not only in planning and restructuring the process, but also in implementing it to ensure that it achieves its intended results. These realities give rise to the following tenets.

Tenet 11: *The psychological and emotional barriers to change must be accounted for and carefully managed throughout the process reengineering project.* This point can hardly be overemphasized. If changes are simply levied from above, entrenchment is inevitable. Furthermore, concerns expressed over the so-called resistance to change may represent a tacit acknowledgment that change originates by decree.

Tenet 12: *Users of the reengineered process need to understand their role in support of the process and be trained accordingly to perform their new responsibilities.* Since radical changes in the process are to be expected, the job responsibilities of the process users will probably be changed significantly as well. It is a grave mistake to assume that everyone will understand what is expected of them and that they will automatically know how to carry out their new job responsibilities.

These 12 tenets deal with a broad array of pivotal issues related to the successful process reengineering project. They also establish the foundation for the process reengineering model described here. But more than this, they speak to the heart and soul of how an organization carries out its mission on a day-to-day basis—not simply how a particular reengineering project is structured. It makes little sense, for instance, to charge a reengineering team with the task of revamping a certain business process in order to maximize its value to the customer if the organization as a whole has a history of indifference toward this outlook. Simply put, any condition that would militate against the success of a reengineering project at its outset will be magnified many times over once the project is under way. Therefore, before embarking on a reengineering project, a company should critically examine its culture in light of these 12 tenets. Some groundwork may be necessary in order to reengineer peoples' attitudes and perceptions before attempting to radically reform any business process.

REFERENCES

1. Ralph E. Steere, Jr., *Office Work Simplification* (Englewood Cliffs, N.J.: Prentice-Hall, 1963). This informative little book offers guidelines that are as relevant today as they were in 1963. The author also lists references on the subject dating as early as 1951. While certain parallels exist between the methods of work simplification and those employed under the rubric of process reengineering, the latter poses significant challenges to such basic assumptions as the intent of the process and the distribution of decision authority—especially in light of today's sociotechnological work environment and the more contemporary emphasis on value as defined from the customer's perspective. These issues highlight the fact that the distinction between work simplification and process reengineering goes beyond semantics.

CHAPTER 3

POSSIBILITIES AND PITFALLS

It is common knowledge that any medicine has side effects; process reengineering is certainly no different. Organizations that choose to undertake a major process overhaul should, at the very least, be aware of the potential risks and rewards from so doing. Moreover, this information can be used by upper management to establish decision points during the reengineering project and then make decisions that weigh the benefits against the risks. Knowledge of the various risk factors can also be used for risk-management planning. This chapter provides information that will be helpful in mitigating risk and in making decisions.

Carrying the medicine analogy a step further, it is also a fact that no single medication can effectively treat every conceivable human malady. Likewise, process reengineering is not a panacea for treating every ill that plagues the various processes within an organization. In addition to highlighting potential risks and rewards, this chapter also provides insight into the most favorable opportunities for improvement. More specific guidelines for identifying and selecting a process to reengineer are addressed in chapter 5.

POSSIBILITIES

Some of the most notable process reengineering successes have occurred simply because someone in a position of authority had the foresight to challenge conventional wisdom by asking certain questions.

- Why does this particular process need to exist versus some alternative solution or no process at all?
- Why does the process require this much time to achieve certain results?
- Why is the process organized this way?
- What aspects of the process are truly important to the customer?

Consider, for example, the Japanese auto manufacturer, Toyota, where the asking of such questions led to radical process improvement.[1] While, up until the 1970s, U.S. car makers operated on the long-standing assumption that it requires six to eight hours to change production runs from one set of specifications to the next, Toyota engineers challenged this assumption and found ways to reduce the change-over time to a matter of minutes. Consequently, Toyota was able to increase the production capacity of its assembly lines, carry smaller inventories, and respond more rapidly to its customers. In other words, Toyota significantly *improved its competitive advantage in this case by reengineering the process rather than the product.* (In the automobile industry, product design is in a perpetual state of evolution among *every* manufacturer, making it more difficult to secure a competitive advantage through product design features alone.)

The assumption among U.S. car makers regarding changeover time was largely predicated on an inventory control model that has existed for over half a century. Apparently, the tenure of this model had something to do with making it a de facto law that few, if any, felt compelled to challenge. While

U.S. car makers have since adopted this new paradigm, Toyota is still able to outperform its U.S. rivals in this regard, at least partly because they have been able to take advantage of their early lead and continue to improve the changeover process.

Even though this case involves a production process, Toyota's willingness to challenge sacrosanct methods and assumptions is directly relevant to the reform of business processes as well. It also underscores the fact that (1) process improvement can be used as a powerful weapon for achieving a competitive advantage and (2) it is usually better to be the instigator of change rather than the follower. Other success stories involving business processes are profiled in chapter 14.

Improved Customer Focus

Few would argue—at least today—with the notion that being customer focused is essential to the survival, and certainly the growth, of the organization. And relying on customer focus as a competitive strategy suggests that an organization is positioning itself to be better than its competitors at translating the customer's desires and needs into cost-effective solutions. It also suggests a close alliance with the customer and an understanding of what he or she truly values, in contrast to the traditional, and impersonal, niche or demographic approach to characterizing market segments.

At the business-process level, the customer's wants and needs are translated into solutions by maximizing those aspects of the process that add value while minimizing those that add cost but comparatively little, if any, value from the customer's point of view. Recall from the previous chapter the first tenet of process reengineering: *The customer, and the customer alone, is responsible for defining what constitutes product or service value.* Thus, a successful process reengineering project will, by definition, focus on the value aspects of the process and ensure that each stage of the process effectively and efficiently generates solutions that match requirements. Also note that in contrast to the quality philosophy that stresses "doing things right," the value orientation emphasizes the importance of *doing the right things right.* Even a flawlessly designed buggy whip will not sell if the customer does not need a buggy whip.

Better Linkage Between Strategy Plans and Business Operations

A recurring frustration among upper management is that strategy plans are so seldom propagated throughout the organization and reflected in business operations. Of course, the finger is often pointed in the other direction as well, as indicated by such statements within the ranks as, If they would only let us know what they want, or If they would simply leave us alone and let us do our jobs.

When successful, process reengineering will have a positive impact on the deployment of strategic plans. Ideally, business processes can be reengineered to include the priorities of the organization among the customer requirements these processes are designed to support—adding to the justification for executive-level involvement in the reengineering project. In addition, the emphasis on trust and communication throughout the project (as articulated by Tenet 8) contributes to improved linkage between strategy plans and business operations.

Dramatic Improvements in Cycle Times

Process cycle time refers to the duration of time it takes for a certain process to run its course from one end to the other. For instance, in processing insurance claims, the cycle time might be the *average* length of time from the point of receiving the claim to the point of closure. If the process conditions change in response to different circumstances, say the processing of liability claims versus accident claims, it may be necessary to establish an average cycle time for each set of circumstances. (Note that in addition to improving the cycle time of the process, we most likely would want to improve the variance in cycle times between process runs as well. It is easier, for instance, to plan a certain project if we know that the purchasing cycle requires 10 days, plus or minus 1 day, rather than 10 days, plus or minus 5 days.)

There are at least a couple of important reasons why improving cycle time might compel an organization to reengineer a business process.

1. Delays in the process, whether they are deemed unavoidable or not, are a major source of discontent to the customer. (In this case, the *customer* is anyone downstream in the process from the point of the delay, whether they be internal or external to the organization.) On the other hand, superior performance in this regard can be used as a powerful competitive weapon.

2. Time is money—especially in the realm of process management. Anything that causes the process cycle time to be longer than absolutely necessary adds avoidable cost to the process. While there are others, sources of delay in processing may include duplication of effort, unnecessary tasks or subprocesses, bureaucratic red tape, and recycling the flow of work between subprocesses. Reductions in process cycle time will ultimately be reflected in lower costs to the customer and higher profits to the organization.

Improvement in cycle time is in essence the aggregate outcome of several measures taken to make the process more efficient. In general, two approaches lead to improving the efficiency of a process.

1. Incremental improvement over time of an existing process, relying, perhaps, on a widely adopted model that has become known as the continuous improvement process (CIP)

2. Radical overhaul and reengineering of the process to achieve dramatic, and hopefully instantaneous, results

These two approaches should be considered allies in the war on waste and inefficiency in the organization. Even a radically overhauled process will require fine tuning as the need for improvement becomes apparent over time. Nevertheless, there are limits to how much can be achieved through incremental improvement alone. Furthermore, some organizations have found that CIP can take on a life of its own, as indicated by the following comments in a *Business Week* feature on the quality imperative.

> *James L. Broadhead, CEO of FPL Group, Florida Power & Light's parent, dismantled the quality-improvement department which had swelled to 75 employees. "It was becoming a bureaucratic organization that was outliving its usefulness," says Michael T. Fraga, an FPL vice president who today heads a quality staff of just six. . . .*
>
> *Alcoa Corp.'s CEO Paul H. O'Neill just scrapped the company's decade-long continuous improvement strategy calling it a "major mistake." Alcoa, he decided, needed "quantum" quality improvements instead to meet world standards.*[2]

While it may be a mistake for an organization to go so far as to scrap its continuous improvement strategy, every organization must closely monitor the efficiency of its core business processes in light of the perpetually changing business environment. The message here is not that process reengineering displaces continuous improvement, rather that with incremental improvement, one should be alert to the danger of perpetuating the status quo, perhaps adding patches and bandages to a process that is in serious need of reform.

Given that the CIP philosophy empowers and depends on those who own the process to identify areas for improvement, it is easy to see that local interests can sometimes prevail against the global interests of the process or the larger interests of the organization. For this reason, radical process reform, at least in situations where it impacts peoples' jobs or the basic structure of the organization, requires direct intervention from upper management or a designated change agent who is not a part of the process itself.

Improved Process Efficiency

Process efficiency is concerned with how well the process uses available resources to achieve the desired results. Since improvements in process efficiency are concerned with the elimination of waste and the need for rework, process efficiency corresponds to the concept of quality as some would define it. Perhaps not surprisingly, there is an inverse relationship between process efficiency and process cycle time, as depicted by the plot in Figure 3.1 for a hypothetical situation.

Figure 3.1 plots the relationship between process efficiency, represented by the Greek letter eta (η), and the cycle time *for a given process configuration*. The dashed lines in this example show how the cycle time decreases from T_1 to T_2 when the efficiency is made to improve incrementally from η_1 to η_2. The curve of the plot also suggests that returns diminish somewhat as the process efficiency is further increased. In other words, it becomes more difficult to achieve improvement in cycle time as incremental improvements continue to be made in the efficiency of a particular process configuration.

Figure 3.2 shows how conditions might change if the process is radically altered. The lower curve, in this case, plots a set of points that corresponds to the new process configuration. From this we can see how cycle time can be improved by radically reforming the process. Notice also the relative degree of incremental improvement in efficiency that would have to be made in the original process to realize the same degree of improvement in cycle time that is achievable by

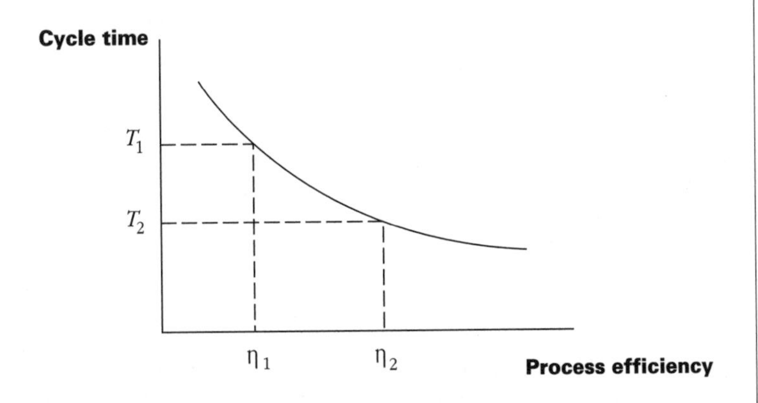

Figure 3.1 Process efficiency versus cycle time.

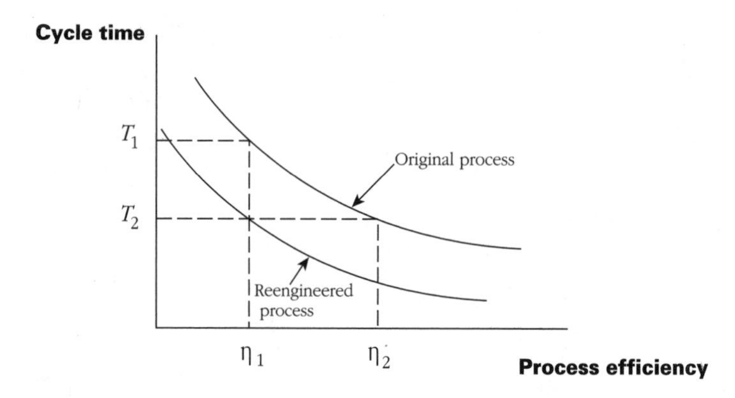

Figure 3.2 Cycle time improvement with reengineered process.

reengineering the process—especially as the efficiency nears its peak for a given configuration. This quantum leap in cycle time improvement is what is meant by the expression *breakthrough success.*

The goal of process reengineering is to transform the process such that the lower curve is driven as far down as possible, thereby realizing a quantum degree of improvement. But just how much improvement is possible even if the process is reengineered? We may never know for sure, but we can break down the process cycle time into the following three factors that are helpful in making this assessment.

- *Real value added (RVA)*—refers to activities that are essential to the process in order to meet the customer's expectations

- *Business value added (BVA)*—refers to activities essential to conducting business, such as policy and regulation compliance, that add cost to the process but do not add value from the customer's perspective

- *No value added (NVA)*—refers to activities that neither add value to the process from the customer's perspective nor are required to conduct business, such as storage, movement, rework, approvals

Cycle time improvement occurs when an organization becomes more efficient at performing the RVA activities (or subprocesses) and eliminates, or at least minimizes, as many of the BVA and NVA activities as possible. Therefore, if the process is analyzed in terms of these three factors, it is possible to obtain some idea of the degree of improvement that is possible. We will have more to say about this in chapter 5.

PITFALLS

Now that we have had a glimpse at some of the rewards that are possible through process reengineering, let's examine some dangers and pitfalls. Most of these can be avoided, or at least managed when they arise, if sufficient planning and preparation go into the project.

In terms of risk-management planning, you should seek three pieces of information in relation to each risk factor:

1. The proper *identification and characterization* of the various risk events

2. The *likelihood,* or probability, that a particular risk event will occur

3. The *impact* on the process, the organization, and the customer should a particular risk event occur

The descriptions below provide a starting point for identifying and characterizing the risk events that pertain to your situation. In assessing the likelihood and potential impact of each event, you may need to create a matrix, such as the partial matrix in Figure 3.3, for comparing the risk factors and making early decisions on where to take action. A detailed analysis, and perhaps a risk management plan, may later be necessary as well.

If the scores you assign to the probability and impact factors are highly subjective or speculative, it may be better to use a low-medium-high scale to rate each factor. Also, unless you feel comfortable saying that a high-impact, low-probability

Possible risk event	Probability (1–5)	Impact (1–5)	Assessment
1. Lack of upper management commitment	4	5	High priority
2. Resistance to radical process reform	3	5	High priority
3. Disruption of service to customer	1	4	Low priority
4. Legal consequences from downsizing	2	4	Moderate priority

Figure 3.3 Risk assessment matrix.

risk event is of equal concern to a low-impact, high-probability risk event, be cautious of the meaning of a composite index derived from the product of probability times impact for each event.

Choosing the Wrong Process to Reform

Since it is not possible—or even desirable—to reform every process radically at the same time, careful thought should be given to where to begin. If there is true commitment to the process reengineering philosophy at the executive level, it generally makes sense to select the first process to reform using the following screening criteria in the order given.

1. A process that has a high likelihood of being reformed
2. A process for which the reengineering effort can produce rapid results
3. A process that, when reengineered, will result in significant benefits to the customer and the organization

While these criteria are important for selecting any process reengineering project, they are especially important the first few times. There are enough difficulties to overcome initially without having to also explain that a failed reengineering project was something other than "another passing management fad." This attitude, referred to as *learned apathy,* is extremely difficult to reverse if it has been reinforced time and again.

Inadequate Change-Management Strategy

This pitfall covers a multitude of sins, most notably the failure to understand and deal with the dynamics of radical change as it affects those who own and control the process. Recognize that when it is successful, the change process typically passes through the three phases shown in Figure 3.4.

Also be aware of the impact that process changes may have on the customer or the continuity of service. Even the faulty perception that something is amiss or things would not be changing is enough to scare off some customers. One way of avoiding this syndrome is to solicit input from key customers early in the concept-planning phase of the project.

Because the magnitude of change is potentially severe, change management has significant importance throughout the entire reengineering project. Be aware that the change-related issues will be different in the concept phase of the project than in the implementation phase. Also note that all resistance to change is not necessarily bad, nor is it always easy to tell the difference between intransigence and a valid expression of concern that deserves attention. Unless change-management expertise exists on staff, it is advisable to seek professional support outside the organization in planning for and managing the social dimension of change.

Rigid Organizational Structure

Recall Tenet 2 from the previous chapter: *The organization should be structured from the top down to support its value-added processes, rather than vice versa.* This does not imply that a physical or functional reorganization needs to occur—though indeed it may. However, a possible change in policy and attitude may be needed to position the business processes as customers of the functional entities. For example, if an existing process is having difficulty groping its way around the

Figure 3.4 Change process model.

organizational structure, it may be pointless to try to reengineer the process. The infrastructure of the organization itself may remain an obstacle to progress.

Structural rigidity is often bolstered by self-imposed policies and practices—many of which were put in place to manage an unskilled labor force. For instance, one persistent complaint that process management experts express is that the standard cost-accounting policies of most organizations are outmoded and regressive, in the sense that they punish, rather than reward, process efficiency. The principal concern is overhead costs, which are typically allocated on the basis of direct labor rather than process inefficiency. A technique known as activity-based cost management (ABC) is being adopted by some organizations to counter this objection by identifying the "drivers" that give rise to overhead costs in a particular business environment.[3]

Recognize also that certain factors that benefit the functional organization can either work for or against the business process, which typically cuts across functional boundaries.[4] Such factors include the following:

- Grouped and focused expertise or specialists
- Well-defined mechanisms for control
- Clear channels of reporting
- Career development paths

Problems in Establishing the Project Team

An early challenge facing the reengineering project is to establish the project team. Even though managing the project team presents its own challenges, laying the foundation for getting down to business to solve a complex problem is especially critical. The project team is discussed in chapter 7. At this point we simply wish to touch upon a couple of pitfalls to watch for during the formation of the team.

Note here that the team is not necessarily a fixed group of individuals throughout the duration of the project. A slightly different mix of skills, abilities, and attributes may be called for as the project evolves. For example, during the concept-development phase, the team may consist of no more than a few members assigned to an executive task force. Nevertheless, at some point early in the project it is important to identify a core set of team members who can, and are committed to, remain with the project from initiation through implementation. Herein lies one of the potential pitfalls: selection and composition of the project team.

Chapter 7 provides additional guidelines, but for now, recognize that the process reengineering project team should consist of a mix of individuals who

- Understand the existing process from end-to-end
- Are able to conceive of what the process could become
- Understand the mission, strategic priorities, and infrastructure of the organization
- Understand the customer and are in tune with what the customer values
- Understand the process reengineering approach and can serve as agents for change

In addition to these collective attributes, every project team members must have the following:

- The time available to commit to the project
- The psychological willingness to commit to change
- Credibility with the other team members and the process owners
- The ability to think creatively
- A positive outlook and demeanor

- An appreciation for the skills, abilities, and viewpoints of others

If these skills, attributes, and attitudes are not represented and/or the necessary commitments cannot be made early on, the process reengineering team is destined to experience difficulties. Also note that it may be wise to include one or more external consultants to round out the team's technical and/or process skills where necessary.

The other significant issue we have already touched on—internal support for the project team. Not only must senior management's point of view be represented on the project team, the highest legitimate authority associated with the process must show continuing support and genuine interest. The reengineering team should not be held back by departmental budgets, interdepartmental turf issues, strong-willed managers, or threats of retaliation. If these assurances cannot be made in the beginning, the project team will not be able to accomplish the desired impact.

Falling into the Technology Trap

Technology is often touted as the engine for change. While few would dispute this assertion, countless examples exist of how technology has actually contributed to the inefficiency of many business processes. The following quotes underscore this reality:

> *Observes Carla Paonessa, a partner at Andersen Consulting. . . "just automating something that shouldn't have been done manually won't get you to be more productive." What will work is eliminating bottlenecks, reducing mistakes, focusing on customer service, and then, and only then, introducing new technology.*[5]

> *Gregory Smith the VP and General Manager (of the California State Automobile Association). . . says he and other CSAA veterans have learned the hard way that "when your processes are a mess and you automate, you get an automated mess."*[6]

Traditional process structures were created before the advent of the computer, and for 40 years we have scarcely disturbed them. We automated what was already in place and missed the real potential of computer technology: to support entirely new models of how work is performed.[7]

We waste as much time making sure the margins or fonts on our documents are flawless as we do making sure the contents and ideas are correct.[8]

When the only tool you have is a hammer, you tend to view all problems as nails.[9]

The message here is clear: technology should serve the process, not vice versa. In other words, technology should be viewed as an enabling mechanism rather than the driving mechanism for change. Take information systems, for instance. Such systems have begun to emerge as an important enabling technology. Nevertheless, a considerable amount of time and capital continues to be wasted on systems that ostensibly have reasons of their own for existing rather than being *strategically embedded* within the business processes that create value from the customer's perspective.

Though automation may indeed enhance the efficiency of a business process, technology-based solutions, if necessary, should be determined after the process has been streamlined and optimized. Be alert to the influences, both internal and external to the organization, that do not respect this order.

Before moving on, we should touch on several other issues that may bear on the success of the reengineering project.

Displaced employees: Radically reforming a business process may very well result in the loss of jobs. Certainly this a serious matter, the ramifications of which should be considered well in advance. Also be aware that employees who even suspect that their jobs are in jeopardy will naturally resist any efforts to reform the process. Even if jobs are preserved, radical changes in business processes can still have significant consequences in terms of loss of prestige, loss of career opportunities, and loss of power, to name a few. Without exception, the success of the reengineering project will hinge on how well the human issues are handled.

Failure to properly measure and map the process: As later discussed, one of the first tasks required of the reengineering project team is to analyze the existing process thoroughly. With complex business processes, this can be an exacting chore. Be prepared to deal with the pressures to shortchange this important task and "simply get on with the job of revamping the process."

Problems with implementation: Without exception, implementation is undoubtedly the most challenging phase of the reengineering project. Problems that deserve special attention during this phase include the tendency to drop the new process on the employees without providing them with adequate training, failure to prepare the organizational infrastructure to support the newly reengineered process, failure to pilot test and work out the bugs in the new process, and failure to adequately plan for phasing in the new and phasing out the old processes. The human issues described above will also be intensified during implementation.

Chronic impatience: Reengineering any process takes time—often more time than senior management is willing to commit in relation to the magnitude of the job. If pressure persists to achieve results quicker than the approved project plan would allow, the reengineering project team should be prepared to rescope the project and explain the consequences to those in authority.

While other dangers and pitfalls are certain to arise, those discussed here are some of the more common and potentially devastating. Being aware of their existence and knowing what to look for are important first steps in isolating and managing the attendant risks in your own process reengineering initiative.

As this chapter has described, process reengineering carries its own set of rewards and risks. For organizations that have not undertaken a reengineering initiative, the best advice is to start small, seek quick and meaningful results, learn from mistakes, and leverage success. As an added piece of advice, pay particular attention to the human issues before, during, and after the project. The technical issues will be easy by comparison. Remember, too, the maxim attributed to Samuel Johnson: "Change is never made without inconvenience, even from worse to better."

REFERENCES

1. Edward J. Russo and Paul J. H. Schoemaker, *Decision Traps: The Ten Barriers to Brilliant Decision-Making and How to Overcome Them* (New York: Simon & Schuster, 1989).

2. Keith H. Hammonds and Gail DeGeorge, "Where Did They Go Wrong?" *Business Week,* the Quality Imperative Issue (25 Oct. 1991): 38.

3. Kevin Kelly, "A Bean-Counter's Best Friend." *Business Week,* the Quality Imperative Issue, 25 Oct. 1991, p. 42.

4. Steven J. Heyer and Reginald Van Lee, "Rewiring the Corporation," *Business Horizons* (May-June 1992): 13.

5. Ronald Henkoff, "Making Your Office More Productive," *Fortune* 123 (25 Feb. 1991): 76.

6. Peter Krass, "A Delicate Balance," *Information Week* (supplement) (5 May 1992): 30.

7. Michael Hammer and James A. Champy, "What Is Reengineering?" *Information Week* (supplement) (5 May 1992): 14.

8. David L. Schnitt, "Reengineering the Organization Using Information Technology," *Journal of Systems Management* 44 (Jan. 1993): 15.

9. Attributed to Mark Twain.

CHAPTER 4

PROCESS REENGINEERING FRAMEWORK

This chapter briefly describes a process reengineering framework to assist those who are considering or planning a process reengineering project. Various elements of the framework are described in more detail in subsequent chapters.

As noted earlier, that there is no generally accepted methodology (or cookbook approach) for process reengineering—and for good reason. Business processes exist in an infinite variety of complexities, structures, and intents. Furthermore, every organization has a unique culture that these processes must operate within, which is defined by such factors as the organization's willingness, or lack thereof, to take risks, embrace change, and reward and empower its employees. All these variables have bearing on the approach that will be used to reengineer a process. Collectively, they make it impossible to effectively force fit every process reengineering situation to a single methodology.

On the other hand, certain tried and proven guidelines can be followed. These are characterized here as a *framework* rather than a *methodology,* since they inherently offer greater flexibility than a canned methodology in adapting to a given situation.

For the record, however, note that some management consultants advocate following a structured methodology for this purpose. Such methodologies are fairly common in certain software development environments. One approach proposed for process reengineering is based on the ISO 9000 series quality standards. (These are primarily a set of manufacturing standards that are available through the American Society for Quality Control as the ASQC Q90 series.) At least as far as business processes and the reengineering of business processes are concerned, the use of a rigidly structured methodology appears to be somewhat contrived and, for this reason, less than desirable. Nevertheless, after having successfully completed a reengineering project, some organizations find it beneficial to document the procedures that were followed and use this as a proprietary methodology that can be applied and adapted as necessary to subsequent reengineering projects. In this instance, the methodology can provide more specific guidelines, such as how the project team will be formed, who needs to approve what, and how the project will be funded.

Before examining the components of the framework, we need to make a couple of points regarding the overall structure. Depending on the magnitude of the reengineering project and the complexity of the process being reengineered, some components of the model can be abbreviated and perhaps combined with others. Furthermore, it may be possible to overlap certain tasks, rather than having them occur in a strictly sequential manner.

With these points in mind, let's examine the process reengineering framework depicted in Figure 4.1. Notice that the "front end" (the top) of the model employs what is known as

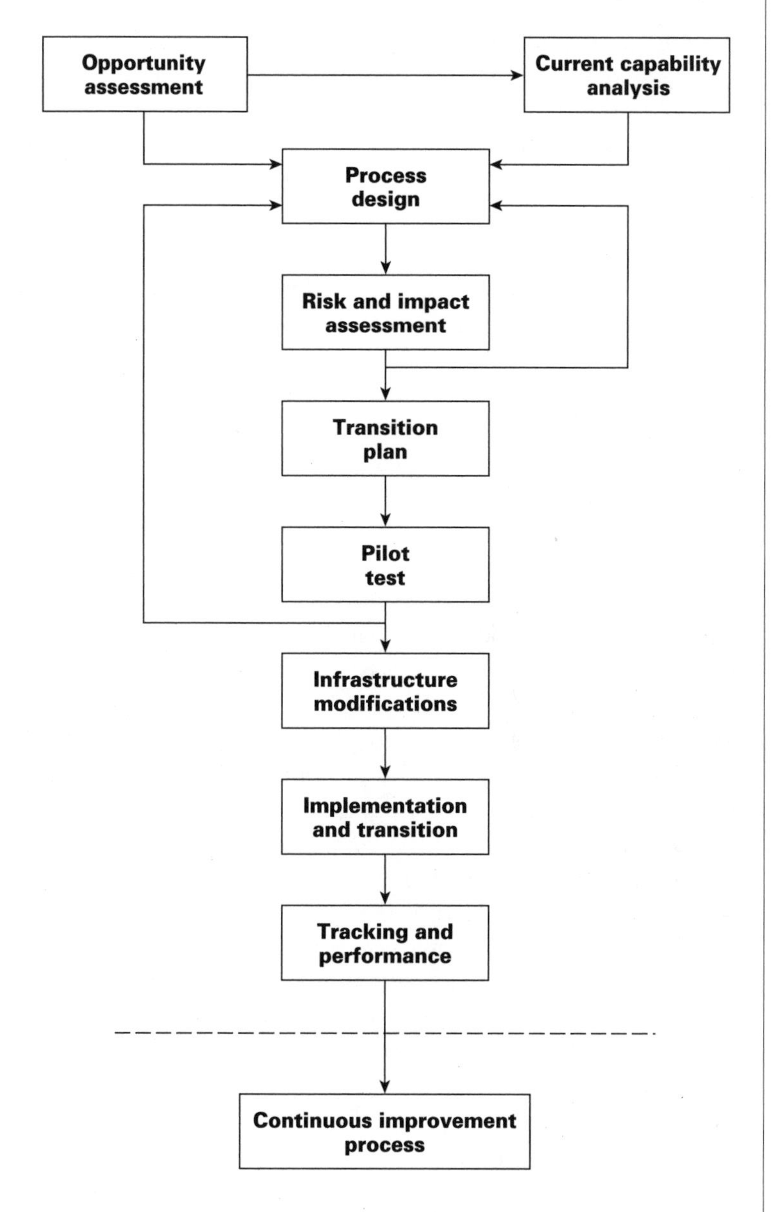

Figure 4.1 Process reengineering framework.

the *gap analysis* technique. Essentially, this technique attempts to determine three kinds of information.

1. The way things should be

2. The way things are

3. How best to reconcile the difference between the way things are and the way they should be

Clearly, measurement and evaluation are critical aspects of the gap analysis. Both quantitative and qualitative data are needed to formulate the conclusions that will be used in defining the process design changes. If the way things should be constitutes only a minor departure from the way things are, then incremental improvements may be called for rather than a radical process overhaul. Often this determination can be made without having to collect and analyze the more extensive data required to reengineer the process.

Also observe from Figure 4.1 the feedback paths that loop back to the process-design stage. While other occasions may require going "back to the drawing board," some design modifications will almost certainly be needed following the risk and impact analysis and, again later, as a result of conducting the pilot test. As with any project, or process for that matter, the later in the cycle a problem is detected, the more costly it is to correct. Research has shown that problems that could have been corrected during the design phase of a project will cost at least 50 to 200 times more to correct once the project is deployed, not to mention costs that can't be quantified.

Technically speaking, the continuous improvement process, shown below the dashed horizontal line, is not a part of the process reengineering framework. Nevertheless, it is shown here to acknowledge the fact that the reengineered process will have to be fine-tuned as problems surface following installation. Such changes are typically identified and implemented by those who own and control the process rather than by the process reengineering project team.

Now that we have seen how the process reengineering framework fits together as a whole, let's examine its components.

OPPORTUNITY ASSESSMENT

The opportunity assessment provides information about the way things should or possibly could, be. We will have more to say about this in chapter 5. For now, note that the opportunity for improvement can be predicated on three primary information-gathering approaches.

1. Systematically analyzing and challenging the assumptions related to the performance of the process, relying primarily on internal sources of information and focusing on waste, inefficiency, and other areas where cost reduction or cycle time improvement can be realized

2. Benchmarking the performance of the process in relation to your chief competitors and/or in relation to similar processes managed by noncompetitors who have demonstrated excellence at what they do

3. Becoming intimately familiar with what your customers value and what motivates them to buy and continue to buy from you or your competitors

Observe that the first two of these information-gathering approaches presumes that you know where you wish to focus your reengineering effort—in other words, which process or processes are candidates for reform. If this is not the case, you will first have to establish priorities in order to make a selection decision. These priorities should, in turn, trace back to the strategic priorities of the organization, unless other compelling factors enter into the selection of the reengineering initiative, such as achieving rapid success.

The third approach is sometimes dubbed "the voice of the customer," most notably when used with a technique called quality function deployment (QFD). While valuable information needed to reform the process can be, and often is, obtained using all three information-gathering strategies, the emphasis on what the customer wants and values should always take center stage. Typically this requires more than a cursory analysis of buying habits or data obtained from opinion surveys. The fact remains, customers may not even be able to articulate what is important to them, especially if something they might assign value to has never been available before. It is also unwise to base process reengineering decisions on casual expressions of interest in which value cannot truly be substantiated.

How, then, can the necessary information be obtained? Some companies have gone so far as to assign a person to work directly with a key customer for a certain period of time, witnessing firsthand how the customer operates while gaining a better appreciation for what he or she truly wants and values. Another approach is to directly involve customers (and suppliers) in the process reengineering effort itself. The better the customer understands the process, the better able he or she will be to provide critical information needed to reform that same process.

Though not strictly a strategy for gathering information, ideas for improvement can also stem from creative thinking. While creative thinking is always important in process reengineering, it becomes especially important if there are no bases for comparison or if customers cannot clearly articulate their needs. As with the Toyota example cited in chapter 3, creative thinking that leads to breakthrough success can get its impetus simply by challenging the multitude of assumptions and practices that perpetuate the status quo. However, ideas for process reform that arise strictly from the imagination also have a downside: greater uncertainty is typically associated with the outcome.

Finally, notice from Figure 4.1 the line connecting the opportunity assessment component and the current capability analysis component. This line is intended to show that the results obtained from the opportunity assessment can be used to narrow the focus of the process reengineering effort and establish priorities, if this has not been done before this point, or if the results of the opportunity assessment shed new light on where you need to focus your effort. In other words, of the many processes that exist within the organization, you need to know which process or processes to target for radical reform. Consequently, the opportunity assessment should not only reveal which customer satisfaction issues are important, it should also indicate each issue's relative importance.

CURRENT CAPABILITY ANALYSIS

In addition to assessing the opportunities, we need to thoroughly understand the capability of the process as it currently exists in order to make rational reengineering decisions. Unfortunately this component of the process sometimes gets shortchanged, perhaps due to overconfidence that we already understand our own business processes and their shortcomings. But several real dangers are associated with this attitude.

1. No one can be truly objective about the performance of a process they have become accustomed to over time. As we do with the windowsill that has forever needed a touch of paint, we tend to become blind to deficiencies that we have assimilated into our current pattern of reality.

2. With complex processes, it is not always easy to tell where one process begins and the other ends. However, it is essential to know what the process boundaries are before we can justify any action that will result in substantial change.

3. Even if a process has been well documented, there may be practical reasons why it is not performing according to design. Employees often devise their own shortcuts and work-around methods to avoid difficulties, confrontations, and bottlenecks that are not covered in the procedures manual.

An important task in performing a current capability analysis is to chart the flow of the process using process flowcharts and possibly data flowcharts and/or geographic flowcharts. The use of flowcharts and other aspects of current capability analysis will be covered in chapter 8.

PROCESS DESIGN

After collecting information suggesting opportunities for improvement and comparing it with information concerning the capabilities of the existing process, we then have a basis for reengineering the process. With this information in hand, the process design component poses the critical questions, What should we be doing that we are not doing, and how should the process in question be modified or enhanced to reconcile the difference?

As Figure 4.1 indicates, reengineering the basic configuration of the process is, itself, an iterative process. When senior management is sufficiently satisfied that the new process configuration does not impose unacceptable risks and that the impact on operations can be managed, it is possible to proceed with planning the transition and pilot testing the reengineered process. Notice that problems identified during the pilot test phase may loop back to the process design stage. While, as stressed before, process design problems must be headed off as early in the project as possible, the design must also be locked in before infrastructure modifications are undertaken to accommodate the refined process. Executive decision points, and perhaps project reviews, can be made to coincide with the culmination of each stage of the process. Elaboration on the process design component is provided in chapter 9.

RISK AND IMPACT ASSESSMENT

As with most situations involving change, many sources of risk are associated with process reengineering. Though it is not possible, or even necessary, to anticipate every potential risk factor, you should recognize in advance where the most severe problems might arise. The complexity and novelty of the project are contributing factors.

Several tools are available for analyzing risk. For instance, a failure mode, effects, and criticality analysis (FMECA) is helpful in identifying and characterizing the various risk factors. A tool known as Pareto analysis can then be used to rank these factors and devise a strategy to manage risk through deflection, mitigation, and/or contingency planning. The use of these tools in a process reengineering context is described in chapter 10.

Although we normally associate risk with negative consequences, at this point in the process we should determine and carefully articulate the potential positive consequences of the reengineering effort as well. Senior management needs to understand both sides of the equation in order to make decisions as to whether to rescope the project or scrap it entirely. While the risk and reward factors will almost certainly need to be assessed in quantitative terms, qualitative data may be all that are available in characterizing such factors as the impact of change on employee morale and customer perception.

TRANSITION PLAN

The transition plan covers details of integrating the process changes into the existing process configuration. Such planning is especially important in a process reengineering context where radical changes in the process can have a significant impact on productivity, continuity of service, finances, facilities, and jobs. Careful consideration needs to be given in advance to such concerns as the following:

- How will changes be made in the infrastructure of the organization to accommodate the reengineered process?

- How will the process changes be phased in to minimize the impact on operations?

- How will those who own and control the process be involved in the transition and trained to use the process in its new configuration?

- What measurements and checkpoints should be established to monitor the performance of the new process configuration?

These and other issues concerned with integrating the reengineered process into the existing organizational structure are further discussed in chapter 11.

PILOT TEST

When properly conducted, the pilot test will provide critical information for working out the bugs in the reengineered process design before the changes are fully deployed. Such tests typically involve some subset of the process workers, the customers, and/or the process that will itself be impacted by the proposed process design changes. A check sheet, such as the one partially shown in Figure 4.2, can be helpful in planning and executing the pilot test.

Depending on the results of the pilot test, it may be necessary to return to the process design phase. Otherwise, minor adjustments can be made without formally reassessing the risks, altering the transition plan, and/or rerunning the pilot test. Either way, the outcome of the pilot test establishes a key go/no-go decision point for senior management.

Also be aware that it is important to highlight which aspects of the reengineered process will not or cannot be

Item No.	Process feature/change	How tested?	Indication of success?
1.1	On-line billing via Electronic Data Interchange (EDI)	Establish EDI links to 2 key customers	Billing cycle delays reduced by 50%
1.2	Consolidated response to billing and technical questions	Train 2 CSRs to handle both billing and technical questions	Customer rating indicating preferred method
1.3			

Figure 4.2 Sample checksheet for pilot test.

addressed by the pilot test. This step requires separating assumptions from facts, including some statement of the degree to which the pilot test is and is not representative of the final implementation.

INFRASTRUCTURE MODIFICATIONS

Two primary issues are involved when considering and implementing the infrastructure modifications.

1. What changes in facilities, technology, systems, and equipment are needed to support the process in its reengineered configuration?

2. How will changes in the infrastructure impact other processes and the organization at large, both during the transition and thereafter?

We have repeatedly stressed the point that one process does not operate in isolation from the others. Consequently, trade-offs and compromises will be necessary to set priorities, often necessitating intervention from senior management. At the very least, contention for internal resources, such as physical space and computer resources, should be anticipated. Also be aware that certain demands placed on the infrastructure should be somewhat lessened when, and if, the process is streamlined as a result of being reengineered.

While cost is always a concern when considering infrastructure modifications, these other considerations are also relevant.

- Where to obtain the necessary technology, systems, and other equipment

- When to schedule the modifications

- How to manage and control the necessary changes in the infrastructure

Typically, these and other concerns that are central to modifying the infrastructure will have been addressed during the risk and impact assessment and elaborated on within the transition plan. At this point in the project, the primary emphasis is on implementing the infrastructure modifications according to plan.

IMPLEMENTATION AND TRANSITION

Once the infrastructure is in place, you can begin implementing the process transformation. Depending on such factors as the magnitude of change and/or the impact on service continuity to the customer, the implementation process may occur in stages, or the existing process may be shut down entirely to shorten the transition time. Such decisions are best made and spelled out well in advance of the actual implementation, preferably as part of the transition plan.

The implementation and transition stage is, without a doubt, the most challenging aspect of the entire project. The principal challenge is to manage the social and psychological dimensions of change. As others have noted, the "soft" issues become the "hard" issues when it comes to process reengineering. Unfortunately, these same issues are often the least understood by those who have the technical know-how to reengineer the process.

If, as stressed before, the process stakeholders have been involved from the beginning, many objections associated with change can be overcome, or at least minimized. Recall from Figure 3.4 that the change process model involves three phases: unfreezing, changing, and refreezing. Each phase of the change model suggests certain context-specific actions that will facilitate the transition. By the time the implementation and transition stage of the project has been reached, the change process should be well into the second phase. Other issues related to the implementation and transition component of the model, such as training, user certification, and change management, will be covered in more detail in chapter 11.

TRACKING AND PERFORMANCE

Once the reengineered process has been deployed, you should validate its performance. In other words, senior management needs proof that the process, in its new configuration, is performing according to plan. For the sake of comparison, the same indexes that were used to measure the performance of the original process will most likely be used to measure its performance in the new configuration. Nevertheless, other parameters may also be monitored if the new process involves tasks that were not performed originally. Also, keep in mind that process performance involves the optimization of a set of subprocesses working in concert to achieve certain performance objectives that have value from the customer's point of view. Furthermore, if the process has been radically altered, you may need to compensate for any learning curve by measuring the performance of the reengineered process after it has to become stabilized.

As with the other deployment components of the reengineering model, plans relating to performance verification should be thought out well in advance and preferably included in the approved transition plan. Unfortunately, this important component of the process reengineering model is often given insufficient consideration. The primary issues relating to the tracking and performance component of the model are covered in more detail in chapter 12.

The framework described in this chapter can be adapted to virtually any reengineering initiative. In essence, it represents a road map depicting and linking the major phases of the project. The framework begins by comparing where we are (current capability analysis) with where we should be (opportunity assessment). If the degree of disparity between the two is large, there may be sufficient justification for reengineering the process. Otherwise, the process can be fine-tuned over time, or perhaps a crossfunctional team can plan and implement fixes that are less intrusive than a radical overhaul.

CHAPTER 5
OPPORTUNITY ASSESSMENT

Every process reengineering project needs some basis for getting started. In other words, the ideas that lead to breakthrough improvement must have their genesis somewhere. In a process reengineering context, this somewhere may be one or all of the following:

1. An effort within your organization to examine a particular process and identify areas of waste and inefficiency

2. An effort to compare, or benchmarking, the performance of your processes in relation to those of your competitors and/or similar processes of noncompetitors[1]

3. An effort to get to know the customer and what he or she truly values about your products or services or those of your chief competitors

In the first case, the standard of comparison is somewhat arbitrary. This review may begin with nothing more than the desire to improve one or more parameters, such as cycle time, without knowing how much improvement is truly possible under the circumstances. For example, upon examining the process involved in filling a customer order, we may find that the process cycle time is 16 hours, on average. Without a basis for comparison, we may arbitrarily set an ambitious goal of reducing this to 10 hours. Quite often, this method of exploring opportunities comes about whenever someone in a key position of authority challenges the status quo. It also stretches our imagination beyond the limits we may have become conditioned to accept—the hallmark of breakthrough thinking.

The benchmarking method of identifying improvement opportunities is predicated on the assumption that an exemplary process, similar in nature to the process under scrutiny, can be identified and examined to establish criteria for excellence. Benchmarking can be accomplished in one of several ways: by studying the methods and end results of your prime competitors; by examining analogous processes of non competitors with a world-class reputation; or by analyzing processes within your own organization that are worthy of being emulated. In any case, the focus for analysis typically relies on one or both of the following:

- *Parametric Analysis*—By this method, characteristics or attributes of similar products or services are examined; price versus performance, for example.

- *Process Analysis*—By this method, the process that serves as a standard for comparison is examined in detail to learn how and why it performs the way it does.

The customer-centered approach to identifying improvement opportunities is predicated on the notion that virtually every aspect of a given business process should be traceable to the customer's wants and needs. Using this approach, it is possible to classify subprocesses, or activities, according to three types of value contributed: real value added (RVA); business value added (BVA); or no value added (NVA) (see chapter 3). When activities are classified in this manner, one can explore ways of becoming more efficient at performing the RVA activities while eliminating, or at least minimizing, as many of the BVA and NVA activities as possible.

This technique is graphically illustrated in Figure 5.1, which plots cost versus time over one complete cycle of a certain process. Figure 5.1 is a cumulative plot, indicating, for instance, that the cost during time interval *T3* is the sum of the costs for activities *A1, A2,* and *A3*. Notice the rate of change at which costs increase for each of the activities as the process progresses over one complete cycle. Using such a plot, it is possible to identify RVA activities that hold the greatest promise for improvement in terms of net gain in efficiency, and the BVA and NVA tasks that add the most overhead to the process. Not only do you want to minimize the rate of increase in cost, but you also want to minimize the time

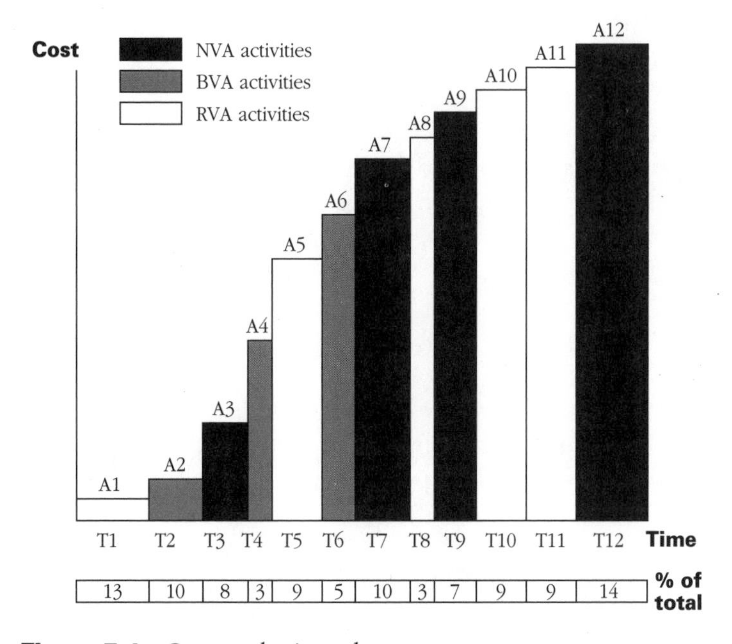

Figure 5.1 Cost–cycle time chart.

required to perform each activity—assuming, that is, the process can be optimized in this manner in relation to other organizational priorities.

Notice from Figure 5.1 that only 43 percent of the total cycle time of this particular process is devoted to RVA activities. Furthermore, by summing the incremental costs associated with the RVA activities and dividing this number by the cumulative cost, it is possible to determine that only some 37 percent of the total process cost, in this example, is devoted to RVA activities. Such percentages are not uncommon for actual business processes. In fact, according to research conducted by a team from Vanderbilt University, the RVA commonly represents less than 10 percent of the total cycle time of the business processes in many industries.[2] Table 5.1 shows the partial results of this research.

Keep in mind that many changes that might significantly improve the cost or time required to execute BVA and NVA activities can be accomplished as part of an ongoing improvement process. If so, this begs the question, Where does continuous improvement end and process reengineering begin? To answer this question, recall from earlier comments that continuous improvement is, as its name reveals, an ongoing process that is normally handled by those who own and control the process. Process reengineering, on the other hand, is coordinated by a specially appointed project team. Furthermore, process reengineering falls under the aegis of upper management for one, or more, of the following reasons.

- The scope of the changes is comparatively large (and costly), possibly reaching beyond the process in question.

- The needed changes could potentially conflict with the personal or professional interests of those who own and control the process.

- The process owners and operators do not have the time, talent, and/or authority to accomplish the needed results in the relatively short period of time that gives rise to most process reengineering projects.

Industry	Process	Average cycle time	RVA time	% RVA time
Life insurance	New policy application	72 hrs.	7 min.	0.16
Commercial banking	Consumer loan	24 hrs.	34 min.	2.36
Hospital	Patient billing	10 days	3 hrs.	3.75
Consumer packaging	New graphic design	18 days	2 hrs.	0.14
Motor vehicle equipment	Financial end-of-month closing	11 days	5 hrs.	5.60

Table 5.1 Cycle times and RVA values for business processes.

Consequently, a process reengineering effort would most likely be called for under the following circumstances.

1. Certain NVA or BVA activities need to be eliminated, or virtually eliminated, from the process.
2. The efficiency of the RVA activities needs to be significantly improved.
3. The entire process needs to be radically altered to achieve optimum results.

Before leaving this chapter, we should stress that regardless of the approach used to identify radical process improvement opportunities, there is a common need to establish objective measurement criteria. Cost and cycle time are only two of many parameters that we may wish to monitor and measure. Depending on the process, we might, for instance, be interested in assessing the following:

- The number of errors introduced throughout the process
- The amount of backlogged work, or work in process

- How many times information recycles between two subprocesses or departments

- The frequency of interruptions in the process

Also be aware that many organizations have benefited by applying the concepts of statistical process control (SPC) to their business processes.[3] SPC offers a formal system for measuring and tracking process performance using basic statistical concepts and charting techniques.[4] The SPC charts provide a visual record of changes in certain key parameters, and as such, serve as a useful tool for identifying actual and impending problem conditions. By indicating process variance over time, the charts may also suggest areas where breakthrough improvements in process performance can be achieved.

Though not currently used on a broad basis outside the manufacturing environment, another statistical technique known as experimental design holds the potential for radically improving certain business processes.[5] This technique is especially useful for optimizing the performance of a process when the process is affected by a number of factors, or variables, that can assume two or more values. Examples of such variables relevant to business processes include overtime hours worked, educational level of the process workers, time spent in training, technology options, and process configurations.

Recognizing the opportunity for breakthrough improvement is a challenge that deserves more than cursory attention on the part of the organization that truly wishes to reinvent itself. Unfortunately, many organizations, perhaps lulled by their financial success in the past, have built a labyrinth of invisible walls that virtually preclude them from visualizing anything better than what they have. This chapter describes some helpful guidelines for removing the blinders that perpetuate status quo thinking.

REFERENCES

1. A number of benchmarking cases have been documented in recent years, most notably Xerox Corporation's initiative to benchmark its logistics operation against that of L. L. Bean. For additional information on this case and the benchmarking method in general, refer to Robert C. Camp, *Benchmarking: The Search for Industry Best Practices That Lead to Superior Performance* (White Plains, N.Y.: Quality Resources and ASQC Quality Press, 1989).

2. Joseph D. Blackburn, "Time-Based Competition: White-Collar Activities," *Business Horizons* 35 (July-Aug. 1992): 96.

3. Cort Dondero, "SPC Hits the Road," *Quality Progress* (Jan. 1991): 43; Steve D. Doherty, "Process Management Has U.S. Air Force Unit Flying Faster," *The Quality Imperative, Profiles of the Human Side* (Sept. 1992): 44.

4. Eugene L. Grant and Richard S. Leavenworth, *Statistical Quality Control,* 6th ed. (New York: McGraw-Hill, 1988); Lon Roberts, *Statistical Process Control for Intuitive Thinkers* (Plano, Tex.: Roberts & Roberts, 1992).

5. William J. Diamond, *Practical Experiment Design for Engineers and Scientists* (Bellmont, Cal.: Lifetime Learning Publications, 1981).

CHAPTER 6

PLANNING THE PROCESS REENGINEERING PROJECT

The point has been made throughout that process reengineering has the characteristics of a project rather than a process, mainly because process reengineering events are episodic rather than repetitive. Rarely are any two process reengineering projects sufficiently alike to generate a "canned" set of procedures that can be applied from one situation to the next. This does not suggest that one cannot benefit from the experience of having executed a process reengineering project. In fact, with experience, one can produce a set of process reengineering guidelines, at least another level of detail below that presented in chapter 4, to incorporate language and structural considerations specific to the organization. These guidelines might even be documented in a "Company XYZ Process Reengineering Handbook." The danger here, of course, is that the handbook will evolve into a rigid set of rules rather than a helpful set of guidelines, defeating the spirit of the reengineering philosophy itself.

Most of our discussion on process reengineering up to this point has focused on the why and the how components of the framework. These issues will be revisited in subsequent chapters, but for now let's direct our attention on the management and execution of the process reengineering project.

WHAT IS PROJECT MANAGEMENT?

In a nutshell, *project management* is concerned with the management of resources in order to achieve the project objectives in the most efficient and effective manner possible.[1] The project objectives, at least in the case of process reengineering, are defined by the process design and the transition plan, or possibly by a process reengineering project plan if the scope and complexity of the project warrants such a plan.

In a traditional sense, project success is measured by the degree to which the project objectives have been achieved, while doing so within a prescribed budget and time frame. These three factors—time, budget, and objectives—are sometimes referred to as the *project constraints,* although the word *constraints* is shunned by some because of its negative connotation. In addition to these three factors, the process reengineering project adds another layer of complexity from a project management point of view: that of skillfully managing the social and psychological dimensions of change. It's a tall order to identify a project team leader who can deal with all these concerns.

THE PROCESS REENGINEERING TEAM

As with other projects, the process reengineering project is executed by a project team. The project team consists of a cadre of people who are assigned to stay with the project, essentially from beginning to end. This does not preclude the possibility that others will be added to the team or removed from it as the need arises. For instance, it may be necessary to acquire one or more technical specialists during the implementation and transition phase. Also, note that being a member of the project team does not suggest a full-time commitment on the part of everyone involved. While the size and complexity of the project will dictate how many people, possessing what skills, are needed to form the project team, at least one person, though typically more, should be assigned to the project full-time. Be aware that team members may not have sufficient time to devote to the project if they have other responsibilities as well.

Setting up and effectively managing the project team is one of the more significant concerns of project management. Chapter 7 has more to say about the process reengineering project team.

PROJECT OBJECTIVES

One persistent problem that plagues the project team is unclear objectives, despite the emphasis in recent years on such techniques as management by objectives and performance-based management. The objectives for a process reengineering project can be particularly elusive, since it is difficult, early on, to pin down the exact nature of the process improvements to be made. Consequently, senior management should play a key role in establishing, and locking in, the objectives for the process reengineering project as early as possible. This responsibility cannot be abdicated, even though the project team may have a role in refining the objectives. It may help in this regard to think of senior management as the customer that is buying the outcome of the process reengineering project. Here, as always, the customer's requirements have to be understood before a service transaction can be consummated.

When they are well defined, the project objectives become a powerful force for building consensus and resolving conflicts within a project team. In other words, the project team will be able to get on with its business much sooner. In most cases, project objectives should focus on outcomes rather than methods. Unless *explicitly* defined elsewhere, a well-stated project objective should include the following:

1. A clear statement of the expected outcome
2. The criterion for judging its success
3. The conditions under which the objectives are to be accomplished

PROJECT LIFE CYCLE

As with other projects, a process reengineering project has a fairly definite beginning and end, and four somewhat distinct project phases in between. In project management parlance, these phases are designated as follows:

- *Concept phase*—The project idea is conceived, research is conducted, the project objectives are defined, and a preliminary project plan is prepared.
- *Development phase*—The design of the product or service that has given rise to the project is developed, and the final project plan is prepared and approved.
- *Implementation phase*—The product or service is brought into existence by being developed, installed, and/or delivered.
- *Termination Phase*—The project is finalized by training the users and making any postinstallation changes, after which the customer buys the product or service in its final form.

For all practical intents and purposes, the process reengineering framework described in chapter 4 has all the characteristics of these four project management phases.

Why be concerned with thinking of a project in terms of its life cycle? Because doing so has several advantages from a planning, tracking, and control standpoint.

- It forces budgets and schedules to be nailed down and formal approval to be clearly given or denied.
- It allows one to anticipate when resources (people, equipment, money, or facilities) will be needed to support the project while under way
- It allows one to define checkpoints, or milestones, where progress can be reviewed and decisions can be made relevant to any changes in scope or design.
- It clarifies to everyone involved that the project is not open-ended and that there are limits to the resources that will be expended.

Also note that the tasks corresponding to each life cycle phase will vary considerably from phase to phase. Consequently, the issues that precipitate conflict within the project team typically change as the project progresses. For instance, priority-related concerns tend to be the dominant cause of conflict during the first two phases, while schedule-related concerns are the dominant source during the last two phases. Anticipating the causes of conflict, and taking measures to avoid them, is an important conflict-management strategy.

Table 6.1 shows the relative rank order of seven of the most common conflict sources as they correspond to each phase of the typical project.[2] Of course, the conflict issues may differ slightly for any given process reengineering project.

Also, in terms of relative intensity overall, conflict issues in the project environment tend to become magnified during the development and implementation phases. But again, this bit of insight and the information shown in Table 6.1 are helpful only to the extent that conflict is anticipated and then prevented from becoming a destructive force. Agreeing at the outset on a fair method for resolving conflict may, for instance, help deal with the conflict issues that are certain to arise.

Table 6.1 Conflict issues by project phases.

Conflict source	Project phase (Ranking sources of conflict)			
	Concept	Development	Implementation	Termination
Budgets and expenditures	7	6	5	2
Interpersonal relations	5	7	5	5
Project administration	2	3	5	7
Technical issues	6	4	2	6
Project staffing	4	5	3	2
Project priorities	1	1	4	4
Schedule/time availability	3	2	1	1

WORK BREAKDOWN STRUCTURE

The work breakdown structure (WBS) is a simple tool for depicting the multitude of tasks that go into each major component of the reengineering project. The WBS can be shown in outline form or pictorially using a hierarchical chart, much like an organizational chart. The following example shows a partial set of tasks for a process reengineering project.

0.0 Purchasing Process Reengineering Project
 1.0 Opportunity Assessment
 2.0 Current Capability Analysis
 3.0 Process Design
  3.1 Develop design objectives.
  3.2 Establish design priorities.
  3.3 Develop revised process flow diagram.

Notice the numbering system used in this example. In this case, three levels of detail are shown. Item 0.0, which simply indicates the project name, stands by itself at the first level, while activities 1.0 through 3.0 share the second level of the WBS hierarchy. Also, tasks 3.1, 3.2, and 3.3 are at the third level, under activity 3.0. If there were any subtasks under task 3.2, these would be numbered 3.2.1, 3.2.2, 3.3.3, and so on. A numbering system such as this makes it convenient to cross-reference the various tasks to such things as work assignments, cost estimates, schedule plans, and narrative descriptions where task elaboration is necessary.

When properly constructed, the work breakdown structure provides a simple yet straightforward means of listing, indexing, and displaying the tasks that go into the reengineering project. Project management software packages are available to help with the mechanics of building a WBS. These tools also allow the user to view the WBS by expanding and collapsing levels on demand. The work breakdown structure offers an excellent starting point for planning the reengineering project.

MILESTONE AND GANTT CHARTS

As shown by the example depicted in Figure 6.1, a Gantt chart is a type of bar chart that shows the project tasks in relation to time schedule. Major milestones are also depicted on this chart.

For planning purposes, it is advisable to begin with a high-level chart, such as the one shown in Figure 6.1, that indicates only the major activities and milestones. Additional Gantt charts and/or milestone charts can then be made that expand the detail to the task or subtask levels if necessary.

Notice from Figure 6.1 how progress toward completion has been represented. A completed milestone is represented by a solid inverted triangle, while the portion of an activity (or task) that is completed is represented by a solid bar. The vertical line shown at the end of week five in this example indicates the current status. As this example shows, the opportunity assessment activity is running one week behind, while the current capability analysis activity is running one week ahead.

While other project scheduling tools exist, such as PERT charts and CPM charts, the Gantt chart is one of the simplest to use and the easiest to understand. The Gantt chart is typically set up after the work breakdown structure has been determined.

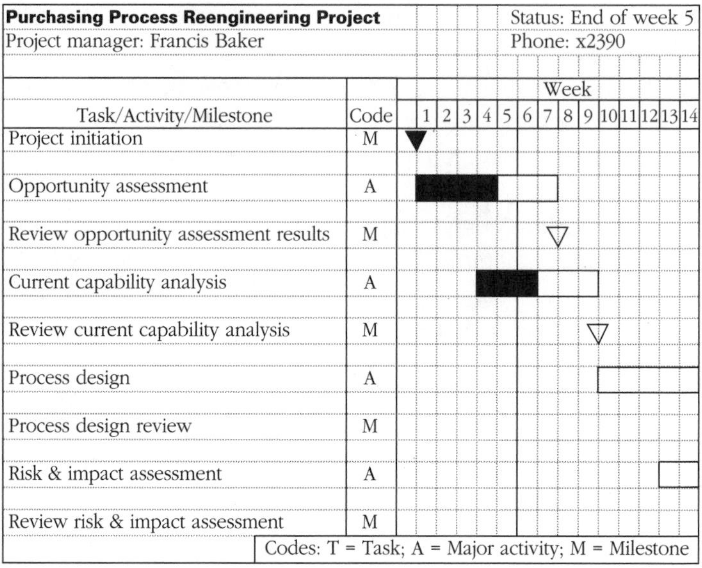

Purchasing Process Reengineering Project Status: End of week 5
Project manager: Francis Baker Phone: x2390

Task/Activity/Milestone	Code
Project initiation	M
Opportunity assessment	A
Review opportunity assessment results	M
Current capability analysis	A
Review current capability analysis	M
Process design	A
Process design review	M
Risk & impact assessment	A
Review risk & impact assessment	M

Week: 1 2 3 4 5 6 7 8 9 10 11 12 13 14

Codes: T = Task; A = Major activity; M = Milestone

Figure 6.1 Gantt chart.

CHANGE MANAGEMENT

The importance of effectively managing change has been stressed throughout this book. It is appropriate to raise this issue again in the context of our discussion on project management.

As noted in chapter 3, a basic change process model consists of three stages: *unfreezing, changing,* and *refreezing.* In terms of planning and managing the reengineering project, it is important to consider the specific tasks or activities that are needed to support each stage of this model. For instance, a specific task related to unfreezing might involve sharing and discussing the results of a competitive analysis with the process owners, thereby encouraging them to visualize the need to radically overhaul their own processes.

Timing, of course, is critical to achieving change and getting people to adopt a new paradigm. Thus, at least some aspect of the change model should be in action throughout the project life cycle. The unfreezing stage may even begin before the project formally gets under way, while the refreezing stage could extend well beyond the installation phase.

Be aware, too, that communication and trust are central themes throughout the change management process. Unfortunately, there are no quick fixes if the organization lacks a history of engendering trust and encouraging open communication. In such cases, the going may be slower, since a negative trend must first be reversed, but the process for achieving change is basically the same.

This chapter has examined some of the issues that need to be considered in managing the reengineering project. As we have seen here and elsewhere, managing such projects is similar in many respects to managing other projects. However, process reengineering projects have these notable differences.

- They must strike a delicate balance between being disorganized and being overly structured, especially during the concept phase, when creative thinking is needed to germinate breakthrough improvements.

- They must be endorsed by and executed under the watchful eye of the executive with the highest legitimate authority over the entities encompassed by the business process being reengineered.

- They require greater attention to the dynamics of organizational change.

- They typically involve a high degree of risk and uncertainty, especially if virgin territory is being explored or if the project impacts the enterprise at large.

The ramifications of these distinctions should be carefully considered by those responsible for planning and managing the reengineering project. Still, this does not obviate the fact that many tools and techniques that have proven their worth in managing more conventional projects are equally useful in managing the reengineering project.

REFERENCES

1. Paul C. Dinsmore, *Human Factors in Project Management* (New York: AMACON, 1990); Jack R. Meredith and Samuel J. Mantel, Jr., *Project Management: A Managerial Approach,* 2d ed. (New York: John Wiley & Sons, 1989).

2. H. J. Thamhain and D. L. Wilemon, "Leadership, Conflict, and Program Management Effectiveness," *Sloan Management Review* 19 (1975): 31.

CHAPTER 7
THE PROCESS REENGINEERING TEAM

Nothing is so critical to the success of a process reengineering project as the selection and organization of the project team. This chapter adds to our previous discussion regarding the process reengineering project team by focusing on the composition of the team and its links to the rest of the organization.

First of all, a disclaimer should be added: A number of factors will determine the size and composition of the process reengineering project team, such as the complexity of the process under consideration, the number of functional entities and individuals who have a role in supporting the process, and the culture of the organization toward dealing with change and innovation. The guidelines discussed here will need to be adapted to take these factors into consideration.

For instance, a low-level process—that is, a process involving only one or two departments and having a relatively specialized reason for existence—may require only three or four team members. On the other hand, a core business process, defined as a process that cuts across many departments and has far-reaching impact on the welfare of the organization, may require 10 or 12 project team members in order to achieve the necessary representation. In the first case, the project team may devote only a portion of its time to the reengineering project, while in the second case, several team members may be assigned to the project full time.

Representational issues aside, large teams tend to lose their effectiveness when issues require face-to-face interaction, most notably those that involve consensus decision making or that otherwise depend on the social dynamics of the team. Nevertheless, with the growing acceptance of such technology-based tools as wireless computers and E-mail systems (electronic messaging systems that can transmit text and images via a wide-area or local-area network), it is becoming easier to link team members or to deploy teams that, due to their size, would be difficult to manage in face-to-face situations.[1] Something akin to an executive committee, representing a subset of a relatively large team, can be appointed to deal with the reengineering issues that require frequent face-to-face interaction.

Also be aware that various specialists may be added to, or removed from, the project team as the need requires. If the project is large and the process changes are complex, one team of specialists may primarily support the implementation phase while another team takes responsibility for the infrastructure modifications. In other words, there can be several satellite teams that are loosely coupled to the primary project team.

EXECUTIVE SPONSORSHIP

Executive sponsorship is critical if a core business process is being overhauled. Even what appear to be lower-level reengineering projects may require executive sponsorship if changes in responsibility or major expenditures are potentially involved. Remember, too, that what appears on the surface to be a low-level process, could, in reality, have hidden ties to the rest of the organization. Without the benefit of a thorough process analysis, it may be difficult to make such an assessment when going into the project. This hidden concern further justifies securing executive sponsorship.

If process reengineering is institutionalized or adopted as a long-term business strategy, it is advisable to establish a governance structure that will ensure executive sponsorship is available when needed. Better still, the executive team should provide leadership in recognizing the need and paving the way for radical process reform. While this role may differ drastically from what some executives are accustomed to, it could be one of the most important steps toward revolutionizing the culture of the organization, preparing it to dynamically adapt to its customers' needs, and setting it apart from the competition.

Executive involvement in the reengineering project should be more than symbolic. The executive should be prepared to commit as much as 25 percent—perhaps more—of his or her time to the project. Some organizations, those that have accepted reengineering as a way of life or see the need for sweeping changes, have gone so far as to appoint a full-time director or vice president of process reengineering.[2] In either case, whether part-time or full-time, the executive sponsor can be expected to fulfill some or all of the following roles.

- Ensuring that the objectives of the reengineering project are aligned with the strategic objectives and priorities of the organization at large (that is, serving as a vertical link to the top)

- Ensuring that interdepartmental turf issues do not hamper the progress of the project (that is, serving as a horizontal bridge that rises above functional walls and barriers within the organization)

- Cutting through bureaucratic red tape that might otherwise hamper the progress of the project or the project team

- Selecting the project team leader and appointing the best available people to the project team

- Following up to ensure that the process in its new configuration is achieving the desired results

Of course, not every executive can fulfill these roles. An executive who has a reputation for being autocratic, indecisive, or afraid to take risks would not be the best choice for the executive sponsor of the reengineering project. Among the many attributes and abilities this individual should possess, the following are especially critical.

- Known and respected within the organization for getting results

- Empathetic to and especially adept at handling human concerns

- Devoid of a personal agenda—that is, not egocentric or precommitted to a particular reengineering solution orientation

- Able to grasp the big picture while being especially attuned to the customer's perspective

- Superior listener

THE PROCESS REENGINEERING PROJECT ORGANIZATION

Before describing a model situation, it should again be stressed that the size and structure of the project organization depend somewhat on the situation. Nevertheless, every process reengineering project organization shares a common set of concerns with regard to the qualifications and attributes of the project team members, whether the team is large or small or the team members work full-time or part-time on the project. These qualifications and attributes were outlined earlier in chapter 3.

However, another principle is universally important to consider when selecting the project team. While the outcome of a process reengineering project often results in improved utilization of information systems, care must be taken in selecting the reengineering team to ensure that it is not dominated by those predisposed to a technological solution. Of course, this same advice applies to anyone who is married to a particular solution as well. As one set of commentators has aptly noted; "Reengineering means helping the business rethink and rebuild the business. It requires people who are business process consultants first and technology experts second."[3]

While the nature of the project may indeed warrant inclusion of one or more information specialists or other technologists on the project team, the team members must be ready, willing, and able to entertain breakthrough improvements that may even diminish the emphasis on automation, if such a move is justified for the good of the process. If a technology-based solution emerges from the project, as discussed earlier, it may later be necessary to assemble a team of specialists to implement that portion of the solution.

With the previous thoughts in mind, let's take a closer look at the composition of the recommended process reengineering project team. A well-rounded team would consist of individuals who can support the various roles identified in Figure 7.1. It is recommended that the team consist of six to twelve members. On a small-scale project, some individuals may have to perform dual roles. A large-scale project may justify having two or more individuals to support a single role.

While not shown in Figure 7.1, the reengineering project team may also receive early guidance from an executive steering committee or an executive council. Once the project team is in place and its charter has been defined, this committee would either disband or focus its attention elsewhere. One of these committee members might remain with the project, for the sake of continuity, and fulfill the role identified in Figure 7.1 as executive sponsor. It is impossible to overstress the importance of securing executive sponsorship and keeping the executive sponsor actively involved in the project. Quoting again the previous commentators: "Reengineering has no constituency within the organization. Its constituency must be found at the top."[4]

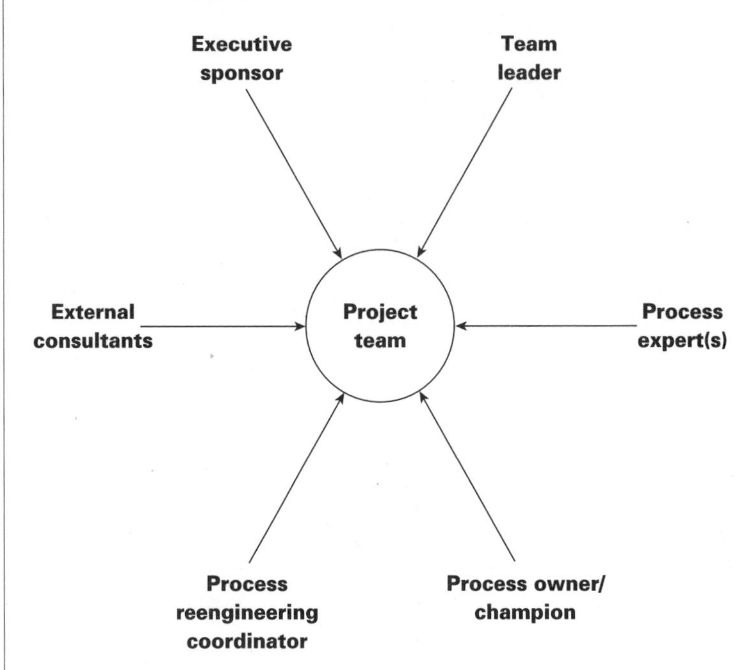

Figure 7.1 Process reengineering project team.

Having earlier defined the role and potential responsibilities of the executive sponsor, let's move ahead and briefly elaborate on the other team positions identified in Figure 7.1.

Team Leader

The team leader essentially fulfills the role of project manager. In most cases, this individual will support the project on a full-time basis. He or she should be skilled at project management and, if not experienced in managing such a project, be trained on par with the reengineering consultant(s) to understand the philosophy and concepts of process reengineering. As a team leader, this individual should be a communicator, an organizer, and also adept at managing interpersonal relationships and resolving conflict when and if it arises. Preferably this individual will enter the project with firsthand knowledge of the process being reengineered.

Process Owner/Champion

A process owner is to a business process what a line manager is to a production line, or a project manager is to a project. In other words, the process owner is an individual who has been given end-to-end responsibility for daily management of the process under study. This person should be interested in ensuring optimum performance of the overall process rather than peak performance of a particular subprocess. But, unless the organization has been structured to accommodate the process management way of doing business, the process owner position, per se, may not exist.

If such a position has not formally been established, an acceptable substitute is someone who can be identified as *process champion*—an individual who has a significant stake in the performance of the overall process. It is not a given, however, that a process champion truly understands the inner workings of the process in question, especially if his or her interest has traditionally been directed toward the outcome of the process.

Process Experts

A process expert is someone who has specialized knowledge of the process being reengineered. While the process expert typically carries out his or her daily responsibilities at the subprocess level, this individual should also have a broad knowledge of the process in question, perhaps by virtue of having worked in several capacities to support the process. This role is especially critical if a process owner does not exist and if a knowledgeable process champion cannot be identified.

The process expert's contribution to the project team is especially important when it comes to collecting and analyzing data relevant to the performance of the process in its current configuration. Nobody understands the existing process better than a process expert who has seen the process in operation day in and day out. This individual will also play a crucial role in implementing the reengineered process and in helping his or her peers understand and accept the proposed changes. Bear in mind that some process experts cannot be objective about the need for radical process reform, especially if they have a personal stake in maintaining the status quo. Depending on the size and complexity of the process, two or more process experts may be needed on the project team.

External Consultants

External consultants are skilled in the reengineering approach to reforming business processes. As a result, they are sometimes referred to as *reengineering practitioners*. These individuals play a key role in ensuring that the reform process stays on track and that problems that might escape those who live with the process in question are identified and handled objectively. The external consultant also serves as a goad by asking probing questions and by being alert to a groupthink mentality that favors premature consensus rather than jeopardize the collegiality of the project team. An external consultant can be independent of the organization but can also be an employee of the organization who has been trained in the process reengineering approach and who is not affiliated with the process being reformed.

Process Reengineering Coordinator

In some organizations, large ones in particular, several reengineering projects may be running simultaneously. The process reengineering coordinator, when such a position has been established, coordinates and integrates the array of reengineering projects that are under way. Some refer to this individual as the process reengineering czar—though the term "czar" has fallen into disfavor due to its gender connotation. This individual works closely with the project team leader, but his or her role is to provide support and global oversight rather than leadership. The reengineering coordinator has a special responsibility for ensuring that the various processes being reengineered mesh with each other and every other process in the organization in an optimum manner.

SELECTING AND PREPARING THE PROJECT TEAM

As mentioned earlier, careful consideration must be given to selecting the reengineering project team. In addition to seeking and selecting team members that can fulfill the roles thus far identified, it is important to consider the "problem-solving compatibility" of the members. In other words, some assurance is needed that everyone on the project team does not share a similar approach to viewing and solving problems.

This advice may seem counterintuitive on the surface, since it is normal to think of *compatibility* in the sense of *getting along*. But, from a process reengineering viewpoint, it is important to select team members who complement one another's problem-solving styles. Furthermore, complex problems—such as radically overhauling a business process—involve varying degrees of divergent (i.e., creative) and convergent (i.e., analytical) thinking. A team consisting entirely of blue-sky thinkers may find it difficult to pin down an acceptable process reengineering solution. Conversely, a team consisting entirely of analytical thinkers can fall victim to analysis paralysis.

Selecting the project team members on the basis of their compatibility is a role ideally suited to the executive sponsor. He or she may be able to make this assessment through personal knowledge of the individuals being considered or by using certain diagnostic tools, such as the Problem Solving Style Assessment[5] instrument or the Myers-Briggs Type Indicator.[6]

Once the project team has been identified, it may also be necessary to provide several briefings, or perhaps more comprehensive training, to prepare the team members to assume their roles. Depending on their individual experience, some may need to better understand the process in question, while others may need training in the fundamentals of process reengineering. In any case, the entire team should be briefed in the beginning by a senior executive as to the nature and intent of the project. This briefing can also serve as a forum for asking questions and airing concerns before leaping into the project.

From what we have discussed in this chapter and elsewhere, several points regarding the selection and contribution of the reengineering project team are worth emphasizing. These include the following:

- The process reengineering team members should be handpicked by senior management based on the degree to which they individually and collectively possess the attributes and abilities described in the current chapter and chapter 3.

- The size of the reengineering project team will be dictated by such factors as the scope of the project and the need to represent the views of the functional entities that support the process. A balance must be struck between achieving fair representation and the potential for diminished team effectiveness.

- The reengineering project will often be shepherded by an executive sponsor who, among other roles, can serve as a direct link to the top and as a bridge for overcoming lower-level bureaucratic barriers.

- Specialists are typically added to the project team, or a satellite team, in cases where input is needed regarding a particular solution or the application of a sophisticated tool. Care should be taken, however, to ensure that specialists do not force a solution that doesn't first, and foremost, work toward streamlining the process and maximizing its value from the customer's point of view.

REFERENCES

1. Tom Peters, "Let the Talk Show Begin," *Forbes ASAP* 151 (7 June 1993): 127.

2. Glen Rifkin, "Reengineering Aetna," *Forbes ASAP* 151 (7 June 1993): 78.

3. Michael Hammer and James A. Champy, "What Is Reengineering?" *Information Week* (supplement) (5 May 1992): 20.

4. Ibid., 24.

5. The Problem Solving Style Assessment™ instrument is distributed through Roberts & Roberts, Plano, Texas.

6. The Myers-Briggs Type Indicator is distributed through Consulting Psychologists Press, Palo Alto, California. Information on an array of psychological tests can be found in such references as Richard C. Sweetland and Daniel J. Keyser, *Tests: A Comprehensive Reference for Assessment in Psychology, Education, and Business* (Austin, Tex.: Pro-Ed, 1990).

CHAPTER 8
PROCESS ANALYSIS

A process analysis is undertaken to help the reengineering project team understand and document the process in its current configuration, or what some refer to as the *as-is conditions.* Such an analysis is the central focus of what was labeled the *current configuration analysis* in Figure 4.1. The process analysis results can also be helpful in assessing the risk and impact of any proposed changes and in helping the project team foresee any difficulties they may encounter during the transition phase.

Though reengineering involves radical process reform, which could potentially be achieved in ways that are not apparent in the beginning, the analysis of the process in question is typically not an open-ended affair. In other words, when a process analysis is undertaken, the effort will be purposely directed toward a specific area, or areas, of concern. The outcome of the opportunity assessment could be helpful in this regard (see chapter 4). Such an assessment is particularly important if it contains information regarding the customer's wants and needs. Lacking specific guidance, however, the analysis process may begin with a gross assessment of the process in question to provide at least some indication of where to target reform. Once this has been established, specific process performance metrics can be identified, and the process can be analyzed in terms of its performance against these metrics.

One caveat should be mentioned before proceeding. While the process analysis results provide information that is essential to process reform, beware of the analysis paralysis trap. Recall from Figure 2.1, for instance, that process flow was depicted at two levels of detail: the subprocess level and the task level. In reality, you may need to break the process into four or five levels of detail if the complexity of the process justifies doing so. Project teams that are enamored of analysis, or perhaps afraid to make a decision, have been known to dissect the process into as many as ten or more layers of detail, typically without accomplishing anything in terms of process reform.

PROCESS PERFORMANCE METRICS

When speaking of process performance metrics, we refer to the indexes we intend to measure in order to evaluate the process in question. Any given process has a number of performance metrics that could potentially be monitored, all with the ultimate intent of reforming the process in relation to one or more of these metrics. Two commonly used process performance metrics have already been discussed: cycle time and process cost. Nevertheless, other performance indicators may be candidates for monitoring.

- Waste, rework, and other indications of inefficiency
- Backlog of work in process, or batching
- Customer response time
- Number of times work is recycled between subprocesses or departments
- Number of document errors or misplaced documents

- Time spent in meetings
- Time spent in transit (for example, transportation delays)
- Customer-satisfaction ratings
- Number of routing errors
- Value-added time
- Number of decision points, including delays in securing approval
- Transcription errors or the frequency with which information must be transcribed

Taking a closer look, it can be seen that some of these metrics are subcomponents of what we might look for when seeking to improve the overall cost and/or cycle time of the process.

Further, we might also wish to monitor certain correlation effects if we believe that one parameter at least partially depends on another. For instance, we may wish to determine the degree of positive or negative correlation, if any, between the backlog of work in process and the time spent in meetings. If we find a strong positive correlation, and other cause-related factors have been accounted for, we could draw the conclusion that by reducing the time spent in meetings, we may reduce the backlog of work in process. Statistical calculations can be used to quantify the degree of correlation between two or more parameters. Simple data plots may also be helpful.

Before leaving this subject, note that the metrics we choose to monitor are also tied to the process design phase of the project. In other words, the process will typically be reengineered to achieve a significant improvement in one or more of the metrics we have chosen to monitor.

MAPPING THE PROCESS FLOW

One of the most important tools available to assist in analyzing and designing a business process is the process flowchart, such as the greatly simplified flowcharts shown in Figure 2.1. Process flowcharts, also known as process flow diagrams, provide a visual map of the way information, documents, materials, and so on are routed through the process. These charts are helpful as both a communications tool and as a tool for pinpointing potential problem areas, such as interface points between functional departments or bottlenecks in the process.

In drawing a process flowchart, we may wish to observe several perspectives. For instance we may be interested in keying in on a certain subset of the process, or perhaps in isolating the end-to-end flow path of one particular component, such as the path taken in processing a check request as part of a larger equipment-acquisition process. We may also be interested in examining how the process flow passes vertically and/or horizontally (that is, crossfunctionally) within the organization. All these various means of mapping the flow of the process may be needed to help us identify areas where the process could be greatly simplified or otherwise radically reformed.

Figure 8.1 depicts the process as it weaves its way throughout the organization. In this particular example, the process has been stratified to indicate how work flows from one group to the next within the organization. The user group represents an internal customer in this example. By organizing the process flowchart in this format, it is possible to see how the work flows from one group to the next as well as one subprocess to the next. This format also brings to light how the workload is distributed across the two work centers and what the impact will be if problems occur within one work center or the other. Of course it is possible, and probably necessary, to carry this process flowchart to one, or more, levels of detail in order to identify problem areas that deserve further consideration. Fortunately, certain software packages are available to assist with the mechanics of drawing, and redrawing, process flowcharts.

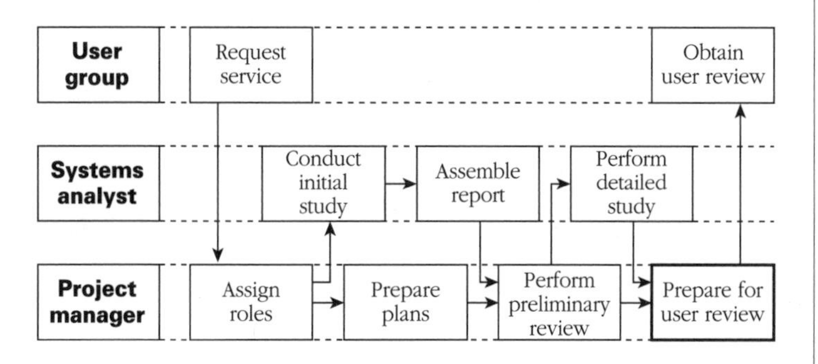

Figure 8.1 Process flowchart.

The format shown in Figure 8.1 can be thought of as a type of matrix that shows task assignments on the vertical axis and process sequence on the horizontal axis. Such a matrix could also be used to show other dimensions of the process. For instance, the vertical axis could be used to depict the three value-added categories—real value added, business value added, and no value added—while the horizontal axis could be used to represent the process sequence, as in Figure 8.1. Such a depiction would allow us to easily identify where value is, and is not, being added to the process.

Certain symbols have become standardized for depicting process flows. For instance, the American National Standards Institute (ANSI) has adopted the symbols shown in Table 8.1.

Other symbol sets are also in use, most of which were originated to depict the flow of data within computerized information systems. The U.S. government, for instance, has adopted the process flow notation known as *IDEF,* which is short for integrated computer-aided manufacturing definition. (Note: IDEF is typically pronounced as an acronym.) Unless you are dealing with a customer that imposes a diagramming convention, such as the U.S. government, there is really nothing sacred about the symbols used to represent the process as long as everyone who needs to understand the flowcharts is clear on what the symbols represent.

Table 8.1 ANSI flowchart symbols.

Symbol	Label	Meaning
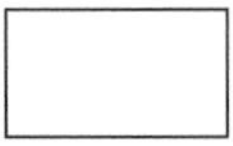	Operation	Depicts a subprocess that receives one or more inputs and then transforms the input(s) in some manner to produce one or more outputs.
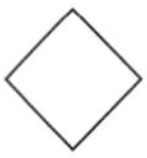	Decision point	A process flowing into this symbol indicates that a decision is necessary in order to determine which one, two, or three paths the process will branch into.
	Paper document	Used when it is necessary to represent a hard-copy document, such as a report, a letter, a requisition form, or a purchase order.
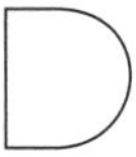	Delay	Depicts a temporary delay in the process, such as when documents are backlogged while waiting approval or when work becomes tied up due to bottlenecks.
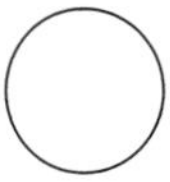	Inspection	Depicts work that has to be inspected or approved before the process can continue past this point.
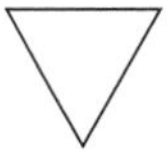	Storage	Indicates where information, or some other work product, is temporarily stored. Some action, such as a job order, may be required to retrieve the work product from storage.
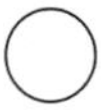	Connector	Depicts connection of points on the flowchart, such as continuation from one chart to the next. A designator is placed inside each symbol-pair to uniquely identify the point.
	Transmission	Depicts the flow of information over great distance, such as an electronic data transmission, a telephone call, or a facsimile transmission.
	Boundaries	Depicts the beginning or end of a certain process. The word *start* or *end* is often written into the symbol.

Preparing of the various process flowcharts needed to map the process in question can be difficult and time-consuming. With complex processes, it is not always easy to determine where one process ends and the other begins. Parallel paths, dead-end paths, and processing exceptions also add to the difficulty of preparing the flowcharts. Furthermore, one operator may opt to route the process along a slightly different path than would another operator. For these reasons, the team's process experts, the people who best understand the process and all its nuances, should primarily be responsible for preparing the process flowcharts. When planning the project, sufficient time—possibly a matter of weeks in the case of a complex process—should be built into the schedule to perform this important task.

COLLECTING AND ANALYZING DATA

Advance consideration should be given to the ways and means that will be employed for collecting and analyzing data. This effort should be factored into the project plan, specifying, where possible, the process parameters to be measured, how the data will be collected and validated, the impact of this effort on the process and the process operators, and the methods used to analyze the data. A time limit should also be placed on the data collection and analysis process in an attempt to avoid the analysis paralysis syndrome discussed earlier.

A number of tools and techniques are available for collecting data relating to current process performance. When possible, the data-collection process should begin by taking advantage of any data that are immediately available. If, for instance, a statistical process control (SPC) system has been in place for some time, reliable, historical data may be readily available.

Since the efficiency of business processes has only recently begun to receive the attention it deserves, the data needed to characterize the as-is performance of the process will most likely have to be collected after the project is under way. If the process cycle time is long or if the performance parameters vary considerably from cycle to cycle, necessitating an average calculation over several cycles, the data collection process can take a considerable amount of time.

In its simplest form, data collection involves the following five steps.

1. Determining the parameters to be monitored
2. Determining how, where, and when the data will be collected
3. Determining how the accuracy of the data will be validated, if necessary
4. Determining who will collect the data
5. Collecting and validating the data

As stated before, to the extent possible, the first four issues should be settled in advance and factored into the project plan. Given sufficient preconsideration, this set of tasks can be accomplished in a fairly straightforward manner. Moreover, data validation may be necessary only if there is reason to believe the data are unreliable or if they have somehow been subjected to measurement bias. Such after-the-fact judgments will likely involve a decision on the part of the executive sponsor. In any case, major process changes should never be predicated on suspect data.

Once the data have been collected, some means is needed of getting it organized and summarized. Data reduction typically consists of relatively simple calculations, such as sums, averages, variances, and perhaps correlation coefficients. One of the most commonly used tools for accomplishing this task is the electronic spreadsheet. Spreadsheets, such as that partially shown in Figure 8.2, can easily be constructed to make the necessary calculations and to prepare summary charts and graphs needed for discussions and presentations.

Taking a closer look at Figure 8.2, notice that each activity on the spreadsheet has been coded using the RVA (real value added), BVA (business value added), and NVA (no

Proposal development process				Date: 7 January 1994
Activity	Code	Process Time (hrs.)	Process Cost ($)	Cumulative Cost ($)
1	BVA	5	600	600
2	RVA	2	300	900
3	NVA	4	450	1350
4	BVA	6	800	2150
5	BVA	2	240	2390
6	RVA	9	900	3290
7	NVA	8	700	3990
8	NVA	16	750	4740
9	RVA	4	270	5010
10	RVA	6	460	5470
11	NVA	7	300	5770
		RVA activities	BVA activities	NVA activities
	Cost:	$1930	$1640	$2200
	% of total:	33.4	28.4	38.1

Figure 8.2 Process performance spreadsheet.

value added) classification system described earlier. In this case, the spreadsheet has been constructed so that the cost associated with each category (RVA, BVA, NVA) can be totaled and then represented as a percentage of the overall total. Armed with this information, we would have a better picture of the cost efficiency of the process. We may also benefit from performing what-if analyses to see how the percentages will shift if the cost associated with certain activities are changed. For instance, if activity 11 is completely eliminated, as a percentage of the new total ($5470), the RVA activities increase by 1.9 percent, the BVA activities increase by 1.6 percent, and the NVA activities decrease by 3.4 percent.

Although it is often painful to reach consensus on how a particular activity should be coded or on the associated time and cost values of each activity, the electronic spreadsheet can be an invaluable tool for organizing, analyzing, and presenting the data once these issues have been ironed out. Given certain enhancements to the example shown in Figure 8.2, here are some additional applications that electronic spreadsheets can help with

- Sorting the records, or rows, by code field to rate the relative importance of the activities within each group

- Creating graphs and charts that can be used for discussion during team meetings and for presentations during project reviews

- Adding a field to display the time variance of each activity and another field for assigning a code corresponding to each activity's flow path so that the total variance of each flow path can be scrutinized

- Testing the impact of across-the-board cuts applied to a particular set or classification of activities

These are only a few of numerous possibilities. The main point is that electronic spreadsheets can be constructed to be as simple or sophisticated as necessary for keying in on those aspects of the process that are deemed important and for making sense out of the data.

In determining the proper classification for a particular activity (RVA, BVA, or NVA), difficulty may arise from attempting to classify an activity that is, in fact, several activities. For instance, if the project team cannot agree on the proper classification of an activity, going one step further with the level of detail may reveal that this activity actually consists of a set of lower-level activities. Ordering materials could, for example, consist of the following lower-level activities.

- Check current inventory level....................................NVA
- Place order...RVA
- Wait for incoming goods...NVA
- Inspect incoming goods...NVA
- Update records..BVA
- Deliver to customer...RVA

In this example, only two of the six lower-level activities were classified as adding real value. A reengineering project team would understandably have difficulty agreeing on the classification of the *order materials* activity when such an activity has elements of all three classifications.

Perhaps from this discussion, it is clear that analysis of the value chain begins with the process outputs and works its way backward, asking at each step along the way, How would the most immediate user (that is, the customer or subsequent downstream subprocess) classify the outputs of the current subprocess? Starting the analysis at the outputs of the process assures us that the links in the value chain are directly traceable to the end customer's requirements and that the requirements imposed by one subprocess on another are not arbitrary.

Structural Analysis

In considering areas where the process in question could be examined for improvement, the discussion so far has mainly centered on the use of certain metrics and measurements relating to cycle time and cycle cost. However, simply being aware of how much time is devoted to, or money is spent on, a certain aspect of the process does not necessarily provide the insight needed to identify the sources of inefficiency. In other words, if the process is inefficient, there are root causes for this condition that can be ferreted out only by being intimately familiar with the process.

The root cause, or causes, of process inefficiency can often be detected by examining the structure of the process. When performing a structural analysis, we are specifically trying to identify missing, overlapping, conflicting, and/or nonessential subprocesses or tasks. This search may lead us to consider some or all of the following:

1. Missing or incomplete management controls
2. Missing quality indicators
3. Missing success criteria
4. Competing or incompatible management priorities
5. Redundant, or unnecessary, information and documents

6. Unnecessary complexity

7. Contradictory information

8. Missing information

9. Valuable process resources committed to making rather than buying products or services that could be obtained more economically elsewhere

10. Employees or process aspects that are overloaded or underloaded in relation to their capacity

11. Unnecessary red tape

12. Conflicting penalties and rewards

These factors deal with structural attributes of the process or its supporting infrastructure. They differ from the variable factors considered earlier, such as cycle time and cycle cost, that focus on the dynamics of what goes on within the process. The term *dynamic analysis* is applied to the latter. Both forms of analysis, structural and dynamic, are important to the process reengineering effort.

PROCESS ANALYSIS IN PERSPECTIVE

Keep in mind that, in the context of process reengineering, the intent of a process analysis is somewhat different than it might be if the goal is to eke out improvements in the existing process. Recall that reengineering does not begin with the assumption that the business process, when reengineered, will even closely resemble its current configuration. Nevertheless, a well-executed process analysis does provide certain information that is critical to the reengineering initiative. It should, for instance, provide insight into the following:

- The *actual performance* of the process in its current configuration. This information can validate the need for reengineering and allow before-and-after comparisons of improvement.

- The *potential performance* of the process in its current configuration. This information can be used to assess the degree of improvement possible without going so far as to reinvent the process.

- The *risk and potential impact* of any proposed changes, especially those that are a radical departure from the existing conditions.

Furthermore, by analyzing the process using the techniques described in this chapter, it is often possible to identify bottlenecks and other inefficiencies that can be remedied immediately. In addition to cutting costs, if properly handled, such rapid improvements can have a beneficial effect on people's perception of the reengineering initiative, perhaps because they eliminate unnecessary tasks that have long been an annoyance or simply improve cross-functional cooperation.

CHAPTER 9

PROCESS DESIGN

There are basically two outlooks regarding process design as it relates to process reengineering. One outlook calls for radical reform by streamlining the process in its current configuration through the elimination of every identifiable source of waste and inefficiency. This approach might suggest eliminating as many NVA and BVA activities as possible while seeking greater efficiencies in the existing RVA activities. By contrast, the *zero base* outlook ascribes to a philosophy of starting with a clean slate and redesigning the process from the ground up rather than trying to revamp what might be a dysfunctional process. With the latter approach, many of the building blocks (or subprocesses) that constitute the current process configuration would naturally be put back into the redesigned process even though the structural configuration of the process—that is, the way in which these blocks interlink and carry out their basic function—could change immensely.

This begs a question: Is it better to pare down an existing process or essentially build up a new one in phoenixlike fashion? The answer, of course, depends on the situation, much the same as does a decision to remodel an existing house instead of building a new one. Factors such as the state of disrepair of the existing process, the impact on the continuity of business and customer service, and the organization's willingness to take risks and deal with change all come into play. Process reengineering, in the strictest sense, is aligned more so with the zero base approach, while continuous process improvement is more closely identified with the other. Since there are limits to the degree that continuous improvement can modify the structural elements of a process, most process reengineering initiatives lean more toward the zero base end of the continuum. As suggested elsewhere, it is the willingness and ability to restructure the process that serve as a significant point of departure between the process reengineering and the continuous process improvement schools of thought.

PRELIMINARY ISSUES

The process design component of the process reengineering framework receives its input from both the opportunity assessment and the current capability analysis components (see Figure 4.1). This suggests that some fundamental issues need to be settled before charging ahead with the design effort. As a minimum, the team should reach closure on the following issues.

- What processes are necessary to fulfill the mission of the organization?
- Do the existing processes fulfill this requirement?
- Does each process meet the basic needs of its customers while supporting the mission and strategic objectives of the organization?
- Which processes should be shut down, reengineered, or continuously improved?
- What, if any, processes need to exist that do not currently exist?

- Can some processes be partially or totally combined to realize greater overall economy and efficiency?

- Where should the reengineering priorities be directed?

These are not trivial issues. It makes little sense, for example, to redesign a process that neither fulfills a basic customer need nor supports the mission of the organization. Unfortunately, such processes are found in most every organization. Assuming from this point forward that suitable answers to these questions can be obtained, let's proceed with our discussion of process design.

PROCESS MODELING

In modeling the process, the emphasis is on laying out a reengineered process structure that will take maximum advantage of what has been learned from conducting the opportunity assessment and the current capability analysis. If, for instance, specific sources of waste and inefficiency have been pinpointed in the current process, certain measures would be taken to ensure that these are eliminated from the process in its reengineered configuration. Specific recommendations and guidelines that will influence the design of the reengineered process will be discussed later, but for now let's focus our attention on a recommended procedure for developing the process model. Keep in mind, however, that if certain portions of the reengineered process must be automated, this should become evident after the structure of the process has been defined. Normally the reengineering initiative does not start with automation in mind. Doing so simply enhances the risk of pursuing misplaced design priorities and perpetuating process inefficiency.

When modeling the process, some advocate the use of a basic building block, such as shown in Figure 9.1. This format reveals the inputs and outputs as well as the control and support mechanism where applicable.

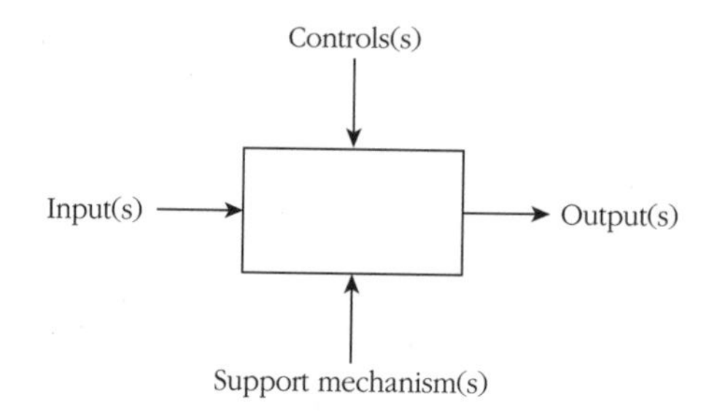

Figure 9.1 Process building block.

A *control mechanism* is anything—policy, procedure, feedback, and so on—that regulates, or somehow modifies, the way in which the subprocess transforms its inputs into outputs. *Support mechanisms* include the human resources, or other resources, that are needed to support the subprocess and to enable it to carry out its basic intent. Of course, not every component of the process will have a control and/or support mechanism. Using the basic building block symbol simply aids in representing the process and in helping to ensure that due consideration is given to the control or support mechanisms, if any, that are associated with the various components of the process. At some point in laying out the process design, other symbols, such as those shown in Table 8.1, may also be needed to provide a full representation of what goes on within the process.

While inputs and outputs normally link one subprocess to another, this is not necessarily true of the lines that depict the control or support mechanisms. For example, a certain set of operating controls—such as a budget limit—can apply exclusively to an isolated portion of the process. In this example, the budget limit is a vital aspect of the process, but, as a control mechanism, it is not a part of the process stream per se. Also, in some cases, as with feedback paths, a control link may be derived or conditioned as a result of passing through some other portion of the process. A system of downstream checks and balances that influences the way the process performs upstream is an example of such a control link. Feedback links are a form of internal control, as opposed to an external control, such as a contract compliance manual.

Typically, when process design is first undertaken, the flow diagram should appear as simple and clean as possible. Details, such as control mechanisms, resource requirements, and decision points can be added after the basic design of the process has been agreed upon. The important point initially is to ensure that all the essential components of the process are represented and that major inputs and outputs are accounted for. In the beginning, it is difficult enough to agree on these factors alone without further complicating matters.

We'll come back to the evolution of the process design, but first let's examine several tools that may prove helpful in synthesizing the process. One tool, the correlation matrix, is helpful in identifying the minimum set of process components. The second tool, the $N \times N$ chart, is helpful in mapping the interrelationship between components. The third, dubbed quality function deployment, offers the advantage of being able to trace each stage of the design back to the customer's requirements.

Correlation Matrix

Often the most perplexing task when designing a process is that of determining which process components are needed, as a minimum, to translate customer requirements into cost-effective solutions. One rather simple tool that has proven effective in this regard is the correlation matrix.

As depicted in Figure 9.2, the correlation matrix offers a simple way of dealing with what is known in engineering parlance as the *covering problem*. In other words, when reengineering a process, we need to ensure that the design is sufficiently complete to cover every aspect of the customer's needs. The correlation matrix does not necessarily ensure that an optimum design has been obtained; rather it ensures that no customer requirement has been overlooked.

In using the correlation matrix, it is best to start by listing the requirements on the vertical axis of the matrix. Process components are then added along the horizontal axis, as necessary,

Requirement	Component/Subprocess												
	A	B	C	D	E	F	G	H	I	J	K	L	M
Respond to inquiries	●												
Acknowledge request for service	○	●											
Customize order to specifications			●										
Provide order-status information				●									
Verify order accuracy		○			●								
Provide billing information	△				○	●							
Verify billing accuracy	△			△	○	○	●						
Charge to customer account					△	△	○	●					
Ship order to customer					△				●				

● = Strong correlation
○ = Moderate correlation
△ = Some correlation

Figure 9.2 Correlation matrix.

to cover each requirement. Check marks can be placed in the cells to show which process component, or components, cover which requirement, or requirements. In this way, it is easy to determine if a requirement has been underrepresented, if process components are redundant, or if certain process components are capable of covering more than a single requirement. Symbols, such as shown in Figure 9.2, can be used in lieu of check marks to indicate how well a particular process component covers a particular requirement. Redundant process components are typically eliminated from further consideration, while those that are capable of fulfilling multiple requirements are retained in the process configuration.

As Figure 9.2 suggests, one approach to dealing with the covering problem is to first assign a component or subprocess to each requirement without giving any thought initially to which components can possibly be combined. This action is represented by the solid dots along the diagonal of the matrix, showing initially that one component has been identified to cover each requirement. Next, each component is examined in relation to each remaining requirement and, where a relationship exists, a "moderate" or "some" correlation assignment is made. At this point we have a better picture of which, if any, components can be consolidated or eliminated from the process design.

In this example, for instance, it may be possible to consolidate the tasks that would be performed by components A and B as well as components E, F, G, and H. Notice that components C and D are critical, since no other components can even partially fulfill the requirement to customize order to specification or provide order-status information. Also notice that it is not necessary to initially provide actual names for each component, unless for some reason we wish to use the components of an existing process as the building blocks for reengineering the process. Component labels (A, B, C, and so on) can be assigned names once the minimum array of process components has been identified.

Note also that the correlation matrix can be used as a process analysis tool. In this case, we would start with the existing process components and then use the matrix to determine how the components correlate with specific customer requirements. Here again any underlaps, overlaps, and/or unnecessary tasks can easily be brought to light.

N x *N* Chart

After the minimum process components have been identified, this information can be used as a starting point for synthesizing the remainder of the process. Input, output, and feedback links between the process components will next have to be determined to establish the relationships between the various components. In this regard, a tool known as the *N* x *N* chart, as shown in Figure 9.3, may prove helpful.[1]

The *N* x *N* chart is a square matrix that shows the main components of the process within the cells along the diagonal. The remaining cells denote the interconnections, if any, between component pairs (that is, any two cells on the diagonal).[2] Comments, symbols, values, and so on, can be added to these cells to describe the nature of the interconnection and/or the work product that passes between the components. Note that outputs are depicted horizontally and inputs vertically. For instance, the cell in the lower left-hand corner of Figure 9.3 intersects component D horizontally and component A vertically, explaining the "D → A" designation. Any notation made in this particular cell uniquely refers to some aspect of the work process that flows out of component D and into component A. If the primary flow sequence is known in advance, it is helpful to list the process

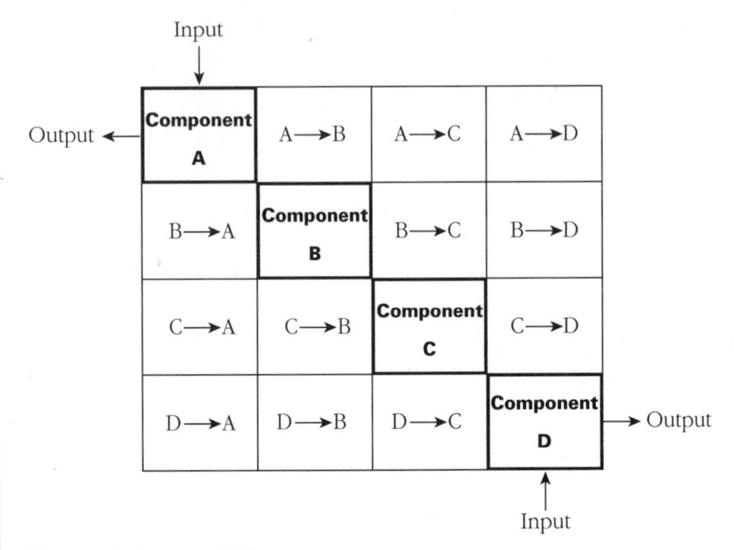

Figure 9.3 *N* x *N* chart.

components in the order they occur, from top to bottom within the diagonal cells. By so doing, the forward-feeding paths will appear to the right of the diagonal and the feedback paths, if any, will appear to the left of the diagonal. This allows any potentially unnecessary feedback paths, which might occur if the work in process is being recycled, to be easily spotted.

Figure 9.4 shows a partial *N* x *N* chart for designing the process described in chapter 1. Recall that this case involved settling insurance claims for the replacement of automobile glass. Note for instance in Figure 9.4 that the customer provides the policy number, describes the damage, and identifies the city nearest to his or her current location. The "acquire claims information" function then verifies the customer data and provides instructions on how the claim will be settled. This same function also provides data to other functions in the process.

As its name implies, the *N* x *N* chart will have N^2 cells, consequently it is sometimes referred to as an N^2 chart. A process that has 12 subprocesses, for instance, would be represented by a 12 x 12-cell matrix containing 144 cells. Since its size increases exponentially with the number of process components, it's easy to see that the *N* x *N* chart can become

Customer	• Policy # • Damage desc. • Nearest city			• Assess satisfaction response	
• Verify data • Give instructions	**Acquire claim information**	• City code	• Claim info.		
• Identify vendor		**Identify local glass vendor**	• Vendor code		• Notify vendor
			Pay local glass vendor	• Vendor code • Pay status	• Issue check
• Assess customer satisfaction				**Update internal records**	
			• Payment discrepancy		**Auto glass vendor**

Figure 9.4 *N* x *N* chart—Processing insurance claims.

unwieldy as design details are added. Nevertheless it is a relatively simple, yet powerful, tool to use. Moreover, the size issue can be somewhat overcome by using one chart to represent the process at the first level of detail and then separate charts to model each major subprocess.

The *N* x *N* chart can be constructed manually or with the assistance of an electronic spreadsheet. In team meetings it may be desirable to use a white board to create the matrix outline and then use adhesive notes to add narrative to the cells. In this way, the cell contents can easily be moved around, added to, or otherwise changed without requiring a complete redo of the matrix.

Quality Function Deployment

With the correlation matrix and the *N* x *N* chart, the task of modeling the process becomes considerably more manageable as we proceed in defining, refining, and adding details to the process design. In this same context, some organizations have found another tool useful. This tool, referred to as quality function deployment (QFD), seeks to develop and deploy the process by ensuring that the *voice of the customer* is propagated throughout the stages that culminate in the final design.[4] Though QFD can vary in how it is set up and what information it seeks to determine, in its simplest form, QFD starts with a clear, customer-oriented objective. From there, a matrix is created showing the degree of correspondence between the *whats* (or requirements) related to the objective and the *hows*, which are essentially the available options for accomplishing the *whats*. This process can be repeated until the details are sufficiently defined, using the most important *hows* identified at one level to represent the *whats* in a subsequent matrix. Figure 9.5 shows a first-level QFD matrix for the insurance claims processing example.

Notice that the objective in this example is stated as a question: What are the important customer response features in settling claims? After a set of *whats* have been identified and listed in the left-hand column of the matrix, the team preparing the matrix completes the task in the following way.

1. They assign a relative importance rating to each of the *whats*. (In this example the ratings range from 1 to 5, where 5 is most important.)

2. They devise a set of *hows* that represent alternatives for accomplishing the *whats*.

3. They rate the degree that each *how* contributes to each *what*. (This is shown as the number in the upper-left-hand corner of each cell in the body of the matrix, here using a 0 to 5 scale.)

4. They calculate the weighted score for each cell by multiplying the corresponding relative importance rating times the raw score. (This is depicted as the number in the lower-right-hand corner of each cell.)

5. They sum the weighted scores for each column to obtain a composite score for each *how*.

6. They determine the relative importance of each *how* by ranking the composite scores.

Objective What are the important customer response features in settling claims?	Relative importance rating	Hows							
		Direct contact w/claims center	One-call service	Phone etiquette training	Established vendor relationships	Involvement of local agents	24-hour hot line service	Mulitple vendors—100 mi. radius	Direct interface w/claims rep.
Minimum time to replace glass	5	5 / 25	5 / 25	0 / 0	4 / 20	0 / 0	3 / 15	2 / 10	5 / 25
Minimum hassle in settling claim	5	4 / 20	5 / 25	3 / 15	4 / 20	0 / 0	2 / 10	3 / 15	5 / 25
Courteous response	4	0 / 0	0 / 0	5 / 20	0 / 0	3 / 12	3 / 12	0 / 0	3 / 12
Reliable glass vendor	4	0 / 0	0 / 0	0 / 0	5 / 20	0 / 0	0 / 0	0 / 0	0 / 0
Dealing with a local agent	1	0 / 0	0 / 0	0 / 0	0 / 0	5 / 5	0 / 0	0 / 0	0 / 0
24-hour service	3	4 / 12	1 / 3	0 / 0	0 / 0	0 / 0	5 / 15	5 / 15	2 / 6
Remote service	2	2 / 4	0 / 0	0 / 0	3 / 6	1 / 2	0 / 0	5 / 10	1 / 2
Speaking to a person vs. a machine	2	0 / 0	0 / 0	3 / 6	0 / 0	5 / 10	0 / 0	0 / 0	5 / 10
	Composite score:	61	53	41	66	29	52	50	80
	Relative importance:	3	4	7	2	8	5	6	1

Figure 9.5 Quality function deployment—processing insurance claims.

In this example, the *how* of directly interfacing with a claims representative was rated the highest overall. Maintaining association with local independent agents was ranked the lowest. The inclusion of the former and the exclusion of the latter are reflected in the reengineered configuration of the process shown in Figure 1.2.

While QFD offers many benefits when properly used, a couple of precautions should be pointed out.

- Every important *what* is tantamount to a requirement that must be covered by at least one how. This suggests that a certain *how* could achieve a low composite score even though it is the only identified solution to a critical requirement. Such a condition might occur, for instance, if the *how* in question is capable of supporting only one requirement.

- In some cases, two or more *hows* correspond positively or negatively to one another. For instance, looking again at Figure 9.5, observe that "involvement of local agents" conflicts with "direct interface with claims representative."

Given the fact that such situations can arise, it is important to have a reality check after completing each matrix. Certain features, such as a correlation matrix, can also be added to help identify any positive and negative relationships between the hows.

By now it should be clear that modeling the process requires attention to details, involvement of others in the design of the process, and most important, taking the customer's perspective into account. While the tools just described can help bring order to this process, they can likewise help identify problem areas that might otherwise go undetected. As described in chapter 13, a number of software tools are also available to assist with the mechanics of planning and modeling the process design. While the latter cannot substitute for human reasoning, they can alleviate some of the tedium involved in organizing information, creating process flowcharts, performing calculations, and making repeated modifications to the design.

Before leaving the subject of process modeling, it should be pointed out that some rather logical, yet often violated, guidelines can be helpful when mapping the process. These are outlined as follows:

1. First agree on the name of the process. Then draw a box depicting the process and its major inputs and outputs. This is sometimes referred to as the concept level, which, as shown in Figure 9.6, is given a 0.0 designator for the insurance claims processing example.

2. Determine the major components, or subprocesses, of the process. A simple listing is all that is necessary initially.

3. Draw a box for each major component and then identify the most significant inputs to, and outputs from, each component. Do not attempt to interconnect the components until all the major inputs and outputs have been accounted for.

4. Interconnect the inputs to, and the outputs from, each major component, adding labels to the interconnecting lines where necessary in order to distinguish one input or output from the other. If, as a result of doing this, you find that a major component is missing, add this component and show its inputs and outputs as well.

5. Repeat steps 2, 3, and 4 by reducing each component, or subprocess, to the next level of detail.

6. Repeat step 5 as necessary until the process is modeled to the degree of detail necessary.

At first glance the concept level depiction of the process may appear rather trivial. Indeed, this is not the case, especially in a reengineering context. For instance, in comparing Figure 9.7

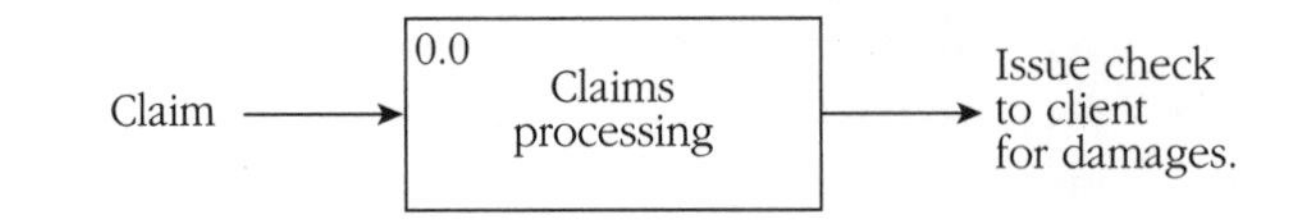

Figure 9.6 Process at concept level.

with Figure 9.6, you can see how the structure of the process would be affected if the emphasis were placed on solving the customer's problem rather than simply issuing a check. Figure 9.7, driven by the voice of the customer, is what gave rise to the reengineered solution shown in Figure 1.2

These steps in the evolution of the process design are rather straightforward. Nevertheless, the dynamics of reaching team consensus while working through these steps can represent a challenge, to say the least. Due to the energy expended and the tension that can result from trying to work through the design process, it is recommended that no more than one or two steps be attempted in any given design session. At the detail level, it may take a number of design sessions to work through a single step. A design session will typically be led by a session moderator—most likely the process reengineering coordinator or an external consultant—and also be supported by a scribe, who is responsible for maintaining a written record of the comments, recommendations, and charts.

SPECIAL DESIGN CONSIDERATIONS

Before moving on, it is worthwhile to consider several important aspects of laying out the process design. In many cases these considerations can inspire creative thinking and prompt people to look at the process in entirely new and unconventional ways.

Poka-Yoke Design

Poka-yoke is a Japanese term that essentially means foolproofing. This concept has been widely adopted in product design and manufacturing, especially in Japan. Many examples come

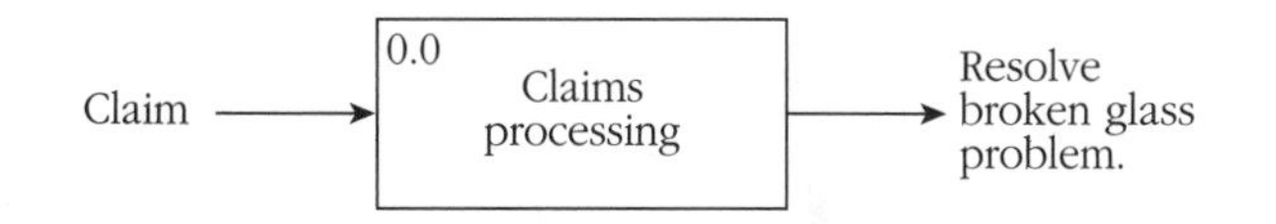

Figure 9.7 Process emphasizing voice of the customer.

to mind, for instance the childproof caps that are standard on medicine bottles and the oversize nozzles on certain gasoline pumps that prevent the operator from inadvertently pumping leaded gasoline into a car requiring unleaded fuel.

In essence, *poka-yoke* contrivances are intended to prevent a problem from occurring in the first place. With a little imagination, it is often possible to identify ways in which business processes can be reengineered to avoid problems. For instance, something as simple as using a color-coding scheme to classify hard-copy files may be helpful in ensuring that files are properly routed.

Robust Design

Robust design attempts to account for the fact that process variation is to be expected, and the process in question should be designed to function normally under such conditions. We expect, for instance, that our automobile will start whether the temperature outside is 10° or 110°F, although at certain temperature extremes, we would not be surprised if it did not start.

As far as business processes are concerned, *robustness* can be engineered into the design of the process in a number of ways. For instance, simply designating a backup person for each key position within a process can keep the process from grinding to a halt when an employee unexpectedly misses work.

Robust design starts with the question, What can cause the process in question to fail in its current configuration? It then seeks solutions that will allow the process to continue without interruption if and when such problems occur.

Value Analysis

Value analysis is a design philosophy that is most usually concerned with the work product. Using value analysis, the process designer would ask, What are the functional attributes of the work product, and how can these attributes be obtained at a lower cost? For example, if a certain work product consists of a written report, from a functionality standpoint, it may be equally effective and less costly to simply place a telephone call rather than create the written report.

In assessing the functionality of a certain work product, value analysis recognizes the existence of two types of functions: work functions and sell functions. *Work functions* relate to the utility of the work product, while *sell functions* are the more subjective features or attributes that make the work product more salable. Both functions should be considered when the work product is under scrutiny. If, for instance, the work product is a certain report, the primary work function may be the report contents, while the sell functions may relate to the color of the cover and the texture of the paper the report is printed on.

PROCESS STREAMLINING

A number of strategies can be employed to streamline a process. However, the inherent danger with streamlining in a traditional sense is that too many compromises can be made in an attempt to keep an existing process intact when, in reality, a radical overhaul of the process is called for. Nevertheless, many strategies listed below may also be appropriate when a process is in need of revamping. It is often the *degree* of change, coupled with executive involvement and the project approach to implementing change, that distinguishes process reengineering from its continuous improvement counterpart.

These strategies are listed below in checklist form (each with a clarifying example) as a simple reminder of other factors you may need to consider when reengineering the process. Some have been described or alluded to elsewhere.

_____Eliminate bureaucracy and red tape.
Example: Empower workers to make decisions that have traditionally required management approval.

_____Eliminate duplication of effort.
Example: Establish shared databases where possible rather than requiring each functional entity to develop and maintain its own independently.

_____Eliminate unnecessary forms, reports, and so on.
Example: Subject every report to the simple test that asks, Who uses this report, and what does it provide that other sources do not provide?

_____Eliminate process bottlenecks.
Example: Establish an alternate processing path to handle exceptions that require a disproportionate amount of time to manage and, thus, tie up the process.

_____Minimize no value added (NVA) and business value added (BVA) tasks.
Example: Examine the basis for company policies that place an extra burden on the business processes.

_____Upgrade job skills.
Example: Ensure that employees have the necessary problem-solving skills and human-interaction skills to function with less supervision, make critical decisions, and handle a broader array of tasks.

_____Simplify the job and the skill-level requirements.
Example: Challenge the assumption that specialists are needed to perform certain tasks by focusing on (1) what the process really requires and (2) the known benefits of motivating workers through greater involvement.

_____Form partnerships with a select number of preferred suppliers.
Example: Establish long-term, mutually beneficial relationships with certain suppliers to leverage purchasing power and ensure consistent quality.

_____Form strategic alliances with other organizations when complementary services can be exchanged for a synergistic effect.
Example: Examine your company's business cycles and consider contracting with external suppliers for any services that are underutilized whenever the cycle passes through a trough.

_____Give employees end-to-end responsibility for a job assignment.
Example: Give employees the authority to make decisions, resolve problems, and interface directly with customers and suppliers.

_____Use automation or mechanization.
Example: Consider using an electronic routing system (also known as work-flow technology*) to support the structural layout of the process and diminish the reliance on forms and other paper documents.*

_____Use standardized methods and procedures.
Example: Establish standards for dealing with problems and train the process workers accordingly.

_____Consider opportunities for make versus buy decisions.
Example: Examine areas of the company that are vertically integrated. Consider buying, rather than making, any products or services that do not benefit from the unique aspects of your organization.

_____Form self-directed work teams that are represented by a cross section of skills and thinking styles.
Example: Assess the problem-solving styles of the cross-functional teams that support each process to ensure that different points of view are well represented when solving problems and making decisions.

_____Adopt a process orientation to doing business throughout the organization.

Example: Define the boundaries of each business process and then designate a process owner to assume end-to-end responsibility.

It is perhaps evident by now that process design is as much an art as a science, maybe even more an art. Open-mindedness and creative thinking are certainly key ingredients. Process reengineering works best when we are willing to challenge the status quo and envision radical ways of transforming and improving the process. This outlook is especially critical during the design phase, since most of what happens beyond this point is concerned with implementing the process in its reengineered configuration.

REFERENCES

1. R. J. Lano, *A Technique for Software and Systems Design* (Amsterdam: North Holland Publishing, 1979); James A. Lacy, *Systems Engineering Management: Achieving Total Quality* (New York: McGraw-Hill, 1992).

2. A square matrix provides exactly the number of cells needed to represent the N components on the diagonal and the complete set of interconnections between any two of these components. This fact can be demonstrated mathematically as representing the number of permutations of N items taken two at a time. Under these conditions, ${}_NP_R = {}_NP_2 = N^2 - N$. Since N^2 represents the total number of cells in the matrix, and N represents the number of components on the diagonal, exactly enough cells remain to represent all possible interconnecting pairs.

3. Lawrence R. Guinta and Nancy C. Praizler, *The QFD Book: The Team Approach to Solving Problems and Satisfying Customers Through Quality Function Deployment* (New York: AMACOM Books, 1993).

CHAPTER 10

RISK AND IMPACT ASSESSMENT

You'll never have all the information you need to make a decision—if you did, it would be a foregone conclusion, not a decision.

David Mahoney
Confessions of a Street-Smart Manager[1]

Ironically, the subject of assessing risks is often glossed over when planning the reengineering project, even though it has never been a secret that reengineering projects often involve considerable risks. Perhaps this is due to our natural tendency as humans to deny the possibility of negative consequences and tacitly believe, If I don't think about it, then it won't happen. Risk assessment is also wrongly perceived by some as negative thinking and consequently avoided. Since it often involves a venture into the unknown, process reengineering on a grand scale is not well suited for organizations that are timid when it comes to dealing with risks head on or reluctant to make decisions largely based on convictions and intuition rather than hard facts.

PROCESS REENGINEERING

Depending on the magnitude of the changes and the complexity of the process in question, there are potentially many factors to consider when assessing the risks involved in reengineering a business process. And as alluded to above, the more pervasive the changes, the less we can know for sure about the success of the outcome. This is all the more reason why a formal risk assessment is critical to the reengineering project. Risk and impact assessment is placed immediately after the process design stage in our process reengineering framework (Figure 4.1) to highlight the fact that the process design options need to be weighed in terms of their potential risk impact as well as their relative merits. If the risks remain too high, it may be necessary to return to the process design stage.

RISK COMPONENTS

Recall from chapter 3 that risk assessment involves three considerations.

1. The proper *identification and characterization* of the various risk events

2. The *likelihood,* or probability, that a particular risk event will occur

3. The *impact* on the process, the organization, and the customer, should a particular risk event occur

Figure 10.1 shows a tool—the risk assessment matrix—that was introduced in chapter 3 for linking these three factors and for assigning values to the probability and the impact associated with each risk event. A simple tool such as this is especially beneficial as a communications vehicle when working in a team setting and when advising senior management on the risks inherent in the reengineering project. If the organization does not have a history upon which to base its risk assessment plans, the pitfall items described in chapter 3 may serve as a

Possible risk event	Probability (1–5)	Impact (1–5)	Assessment

Figure 10.1 Risk assessment matrix.

starting point for identifying and characterizing the potential reengineering risk events.

In addition to the factors cited above, a fourth factor must sometimes be considered when assessing risks: exposure. While exposure is related to the impact of the risk event should it occur, and is sometimes taken to mean the same thing as impact, technically the term refers to the organization's ability to recover from a risk event. *Impact* says, "Here's what's at stake," while *exposure* says, "Here's how we can cut our losses." In a process reengineering context, risk exposure can refer to such factors as the ability to restore operations in the event of a disruption of service, to regain customer confidence if the reengineered process proves untenable, or to recover costs if the process does not perform as expected.

Other considerations being equal, senior management will be more inclined to take risks in situations where the exposure is less. While the assessment of exposure often involves a qualitative or gut-level response, it can sometimes be the deciding factor between two courses of action. Potential exposure at least deserves cursory attention when weighing and presenting various process reengineering options.

Keep in mind that risk assessment is not a one-time affair that occurs early in the project and then is put on a shelf. In forecasting the weather, short-term predictions are always more reliable than long-term predictions. So too with forecasting risks. Therefore, as the project moves along, think continually about the potential consequences of an impending action based on conditions as they exist at the time rather than how they were earlier anticipated to be.

RISK MANAGEMENT STRATEGIES

Advance consideration should be given to the ways in which risks will be managed. Risk management is first concerned with preventing a particular risk event. But it is equally concerned with how to minimize the impact should the risk event occur.

There are three basic planning strategies for managing risk in a project environment: mitigation, deflection, and contingency planning. One, or a combination, of these strategies may be employed to deal with risk. Notice that these are planning strategies rather than action responses.

Mitigation

Mitigation refers to measures that are taken to lessen the probability and/or impact of a potential risk event, typically by backing off on the scope of the project. In a process reengineering context, risk mitigation may mean revamping a smaller number of processes or reengineering some subset of the original process rather than the entire process.

Deflection

Deflection refers to any action taken to divert a risk factor away from the reengineering project by transferring it elsewhere. The risk factor does not necessarily go away. Instead it becomes the responsibility of someone other than the process reengineering project team. Risks are often deflected through negotiation with third parties who agree, perhaps on a contractual basis, to accept the risk as their own and to respond accordingly if the need arises.

Contingency Planning

Contingency planning involves the establishment of options to minimize the impact of a risk event should it occur. While the probability and impact of the risk event may remain unchanged, contingency planning identifies alternative courses of action that may be taken if the occasion arises. For example, if internal support is not available when needed, a software developer could be secured to adapt an existing system to the system in its reengineered configuration. Adding slack time in the project schedule to account for potential delays in the delivery of computer equipment is another form of contingency planning.

REGRETS AND OPPORTUNITY LOSSES

Technically, the subject of opportunity losses is related to cost assessment in general, though it is seldom given the consideration it deserves—perhaps because such costs can be difficult to quantify. What deserves emphasis here, in the context of risk and impact assessment, are the potential negative consequences that accompany *not* selecting a particular reengineering alternative. And, negative consequences involving uncertainty are tantamount to risks, though in this sense, the risks result from not selecting a particular course of action. Such risks are referred to as *opportunity losses*.

Regrets, on the other hand, refer to the losses that can occur when one reengineering solution has been chosen over the other. Consider, for example, the case involving the insurance claims processing center described in chapter 1. In this case, the CEO rejected at least one alternative: retaining the status quo. Though a preponderance of benefits favored the reengineered process configuration in this example, it is conceivable, nonetheless, that a client could have strong feelings about replacing his or her local independent insurance agent with a stranger on the other end of the phone. If customers did react this way, this would represent an opportunity loss to the company, since this particular alternative was not selected. The opportunity loss in this case may not be quantifiable, but it could be significant all the same.

Therefore, in the same sense that it is important to assess the downside risks associated with taking a certain course of action, in other words, analyzing the potential regrets, it is equally important to consider the loss of opportunity that might result from forgoing the selection of other alternatives, including the status quo.

FAILURE ANALYSIS

Failure analysis is a proactive approach to considering the ways in which the reengineering project can experience problems. It usually focuses on major problems that are likely to occur during the implementation and transition phase.

Failure analysis adds another dimension to the assessment of risk by examining the chain of events that could potentially lead to a particular risk event. Ultimately, the intent is to identify root causes of potential problems and take proactive measures where possible to prevent the problem from occurring.

A technique known as failure mode, effects, and criticality analysis (FMECA) is sometimes used to analyze failure conditions.[2] This technique consists of the following steps.

1. Identify the points within the process where major problems might occur or weaknesses might exist.

2. Determine the nature of each failure in terms of what could go wrong (that is, the *failure mode*).

3. Establish what happens to the process or the customer if a certain failure mode occurs (that is, the *failure effects*).

4. Isolate the root cause of the problem.

5. Assess the criticality of the failure.

A criticality index can be established to support the last step by determining the product of the probability and severity (or impact) indexes. A third factor, known as *detectability,* is sometimes used as well in establishing the criticality index for each failure condition. Detectability, in this context, may be better thought of as *nondetectability,* since this index relates to the degree of difficulty in being able to detect a certain failure mode before it occurs. The basis for a detectability index follows this thinking: a problem remaining below the threshold of detection is potentially more serious than one that can be detected before causing even greater problems.

For ease of representation, FMECA can be formatted to fit into a matrix similar to that shown in Figure 10.2. Priorities for taking preventive action can then be established by comparing the relative criticality index of each failure condition.

Notice that Figure 10.2 shows a partially completed FMECA chart for analyzing the failure modes and effects that could occur in the billing function of a certain process. Using this chart, the probability index (P), severity index (S), and detectability index (D) can each assume a value between 1 and 5, though, in practice, no hard-and-fast rules dictate the range of values that each index can assume. The criticality index (C) is simply calculated as the product of P, S, and D. Also notice that the loss of a client, associated with the overbilling failure, was determined to have the highest criticality index from among those shown in this example. Thus, preventing this particular effect would take precedence.

One precaution should be mentioned when interpreting the criticality index. As when interpreting the composite scores in a quality function deployment matrix (chapter 9), a reality check is necessary before taking action on the outcome of the criticality index. This is necessary since the criticality index is computed from P, S, and D index values that may, themselves, be based on subjective judgment. Also, care should be taken to ensure that the evaluators are all rating the same factors. For instance, P and S could logically be associated with the probability and the severity of a particular effect, while D could be associated with the detectability of the failure mode leading to this effect.

PUTTING THINGS IN PERSPECTIVE

The subject of risk has received a lot of attention in the process reengineering arena. Depending on the position a

Subprocess/ Function	Possible failure mode	Possible cause of failure	Likely effect	P 1–5	S 1–5	D 1–5	C PxSxD
Billing	Wrong account number recorded	• Client error • Data entry error	• Late billing • Receivables delayed • Wrong client billed	5 3 3	1 3 5	4 3 3	20 27 45
Billing	Overbilling	• Quotation error • Data entry error • Sales tax added erroneously	• Loss of a client • Errors recorded in ledger • Errors in monthly reports	3 5 5	5 1 1	4 2 1	60 10 5
Billing							
Billing							
Billing							

Figure 10.2 Failure mode, effects, and criticality analysis (FMECA) chart.

certain apologist is trying to defend, one of two extremes is often stressed: the overwhelming payoffs, on the one extreme, or the fear, folly, and failures some organizations have experienced on the other. One writer, for instance, emphasizes the point that "by one estimate, between 50% and 70% of reengineering efforts fail to achieve the goals set for them."[3] Another points to a proprietary study that found that "fundamentally changing a business process produced an average improvement of 48% in cost, 80% in time, and 60% in defects."[4] While these are interesting statistics, in reality, highlighting one extreme or the other is not especially productive—not, that is, for making judgments about the risks and rewards associated with a particular reengineering project.

This chapter has discussed certain tools and guidelines to help the reengineering team assess the risks and impacts associated with their project by carefully considering the factors that are unique to their situation.

REFERENCES

1. David Mahoney, *Confessions of a Street-Smart Manager* (New York: Simon & Schuster, 1988), p. 156.

2. Chris Hakes, ed., *Total Quality Management: The Key to Business Improvement* (London: Chapman & Hall, 1991).

3. Thomas A. Stewart, "Reengineering: The Hot New Managing Tool," *Fortune* 128 (23 Aug. 1993): 41.

4. Clifton Wilder, "Measuring the Payoff from Re-engineering," *Computerworld* (18 Nov. 1991): 65.

CHAPTER 11

PLANNING AND IMPLEMENTING THE TRANSITION

The implementation and transition phase of a process reengineering project is typically the most challenging aspect of the entire project. Unlike continuous improvement, where incremental changes are made to a relatively stable process over a long period of time, the process reengineering project can have an extremely disruptive effect on operations if special care is not taken. Furthermore, continuous improvement initiatives are typically driven by those who own and operate the process, while reengineering initiatives are initiated by senior management. Consequently, the greatest challenge during the implementation and transition phase is almost always associated with the human dimension of change.

The primary purpose of the present chapter is to identify and describe the more important considerations in planning and implementing the transition from an existing process configuration to the reengineered process configuration. Some of these factors were mentioned when we examined the process reengineering framework in chapter 4.

TRANSITION PLANNING

Once the reengineered process configuration has been selected and the risks have been weighed, the task of creating a formal transition plan can begin in earnest. The transition plan should be endorsed by senior management before proceeding with the transition.

There is no standardized format for how to structure the transition plan. Nor are the content areas to be addressed cast in stone. The circumstances will always dictate the format and contents. Nevertheless, the following format may provide a starting point that can be embellished, diminished, or reorganized as necessary:

1.0 Functional Changes

This section describes the changes, if any, in the way the existing process functions. Here it is important to spell out any fundamental changes in what the process does, including any changes in the inputs that will feed the process and/or any changes in the process outputs.

2.0 Environmental Impact

If changes in the process are anticipated to have an effect on other processes, the organization at large, or the products or service delivered to the customer, these impacts should be clearly identified in this section. Here it is also important to elaborate on any changes in the interface points between processes, especially if responsibilities will shift between processes.

3.0 Process Configuration Changes

Any changes in the configuration or organization of the existing process should be described in this section. Before-and-after process flowcharts should be included in this section as well.

4.0 Human Resource Changes

This section elaborates on the effect the reengineered process will have on the process owners and operators. Key positions and changes in responsibility should be clearly identified. If any training and/or certification is required, these should be called out as well.

5.0 Infrastructure Modifications

This section describes any changes in facilities, technology, systems, and/or equipment that are needed to support the process in its reengineered configuration and how these changes will be accomplished. If not detailed in Section 2.0 (Environmental Impact), the impact of these infrastructural changes on other processes and the organization at large should also be described.

6.0 Implementation Phase-In

The reengineered process will often be phased in over some finite period of time, typically to minimize disruption of service to the customer or to stay within a certain budget. If this is to occur, the details of the phase-in process should be described in this section.

7.0 Process Qualification

Process qualification refers to actions, if any, that may be taken to verify that the process, in its reconfigured format, is performing according to certain desired specifications. The performance specifications, the performance measurements, the checkpoints, and the qualification procedure should all be described in this section.

8.0 Transition Team

Depending on the nature of the changes, one or more special transition teams may be needed to support the implementation. Technical specialists, for instance, may be required to reconfigure a computer network, or training specialists may be needed to train and certify the process operators. Any such requirements should be clearly described in this section.

9.0 Timetable

This section of the plan lays out the timetable for getting the transition accomplished. The major milestones, as well as the time duration for each major task, should be identified, perhaps with the aid of a Gantt chart.

10.0 Change Management

The change-management strategy should explain how buy-in on the part of the process operators will be accomplished. This section should also describe any specific measures—such as briefings, Q&A sessions, newsletters, and so on—that will be used to secure acceptance of change. Finally, the section should spell out any special difficulties that may be encountered along the way and how these difficulties will be managed.

Because of its importance, the transition plan deserves careful consideration. Naturally, certain details will not be available until later; nevertheless, the plan should be sufficiently descriptive and comprehensive to make it clear to everyone involved what has to be done to make the transition a success.

TRANSITION IMPLEMENTATION

As the discussion on the transition plan suggests, numerous tasks have to be attended to during the transition phase of the project. Some can be carried out simultaneously to compress the time required to complete the implementation process. Nevertheless, many of these tasks are interdependent, adding to the complexity of coordinating the transition and keeping the project on track.

The remainder of this chapter addresses three of the more critical transition issues: the implementation phase in, the change-management strategy, and the impact of the transition on people and jobs.

Implementation Phase-In

If the changes to the existing process are extensive, these changes will probably need to be phased in over some finite period of time. If so, as a rule of thumb, it is best to *implement those changes first that will realize the greatest benefits.* Early pay-off will go a long way toward building everyone's confidence that the reengineering project will succeed and toward reinforcing the buy in needed to implement the remaining changes. It is difficult for anyone to argue with results.

To do this we need to determine the following:

1. How the changes can be grouped or classified to identify those that can, and possibly must, be implemented simultaneously

2. How much cycle-time reduction and/or cost reduction (or perhaps some other benefit criterion) can be gained by implementing a set of changes

3. How much risk is involved in implementing each set of changes, using whatever risk criteria are most important to the process, the organization, and the customer

With this information in hand, we can then reach a decision on where to begin the implementation process by considering trade-offs in benefits versus risks. A simple matrix, such as that shown in Figure 11.1, can help in this regard.

A matrix such as this makes it relatively easy to compare actions in terms of their risks and benefits. The actual size of the matrix will depend on the degree of detail in which we wish to break out the risk and/or benefit factors. While we would almost always choose to implement any changes that fall within the lower, right-most cell (that is, low risk and high benefit), changes that fall within the remaining cells require greater consideration. Ultimately, any change decisions will depend on the propensity of those in authority to take or avoid risks.

Another factor that may enter into any phase-in decisions is the time required to achieve the expected results. Other considerations being equal, we would normally choose to implement those changes that produce the quickest results, for the reasons cited earlier.

Cycle-time reduction

	0–5 hrs	5–10 hrs	>10 hrs
High risk		4. Consolidate databases.	3. Combine activities E,F, and G.
Medium risk	6. Eliminate activity C. 7. Decentralize buyer responsibilities.	5. Combine activities A and B.	2. Eliminate duplication of effort.
Low risk		1. Eliminate unnecessary forms.	

Figure 11.1 Phase-in decision matrix.

Using a tool such as this, it is possible to see how the phase-in process could extend over several stages. With each consecutive stage, the risks become higher, while the payoffs begin to diminish.

Change-Management Strategy

Acceptance of change is not something that simply happens, even after the reengineered process has clearly demonstrated its worth. Furthermore, nothing can have a greater impact on the ultimate success of the reengineering project than the resistance to change.

Even when the spirit of teamwork is high, people do things for their own reasons. If change threatens something the individual personally values—whether it be authority, responsibility, job security, respect, or even association with peers—resistance to change can be expected. The challenge is to recognize where the process stakeholders are positioned on the denial-to-commitment continuum and, if possible, help them understand how the changes relate to the things that they value most. This is not simply a matter of trying to sell change. Rather, the manager attempts to *enroll* the stakeholders by both inspiring them to ask and helping them answer the question, Why is this important to me?

Clearly change management requires careful consideration and specific actions or, in essence, a strategy. Attention should be paid to the forces that work for and against change. If possible, champions of the changes you are endeavoring to implement should be identified and enlisted as change agents to help lead others past their fears and reservations. This is especially important if these change agents are their peers.

When formulating a change-management strategy, it helps to have a model for change in mind. While a number of change models exist, the change process model described in chapter 3 is relatively easy to understand and especially apropos to the dynamics of change in process reengineering situations.

Recall that this model consists of three phases: unfreezing, changing, and refreezing. In essence, it is concerned with a fundamental change in outlook, or in other words, what goes on within the individual's mind to change his or her perception of reality.

Unfreezing is typically a matter of making the individual aware of the need for change. Awareness is an important tool for guiding people past the biases and mind-sets that lock them into the status quo. In a process reengineering context, for example, the CEO could hold a frank and open dialogue regarding the realities of the marketplace and what is known about the competition.

The *changing phase* of the model consists of helping individuals understand and come to appreciate the nature of the changes that are about to occur. In this regard, a concept known as force field analysis may prove helpful. As Figure 11.2 shows, in performing a *force field analysis,* we seek to determine the forces that work for and against change. Moving from the present state to the desired state may consist of a combination of actions designed to reinforce the driving forces and diminish the restraining forces. However, as a general rule, people are more inclined to accept change if someone removes the barriers and obstacles to change than if someone attempts to

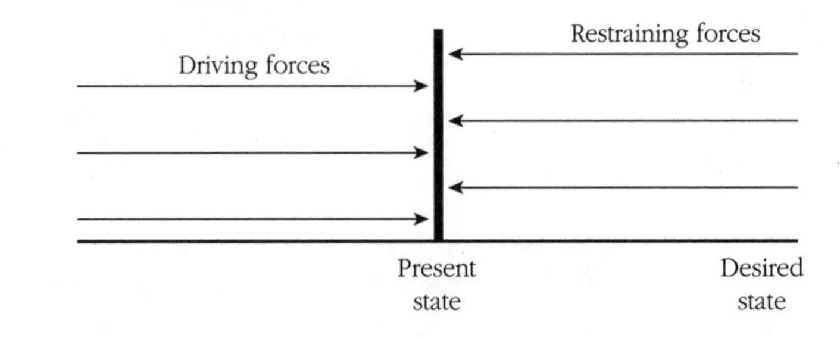

Figure 11.2 Force field analysis.

sell them on its benefits. It would be hard, for instance, for people to believe that a change will be good for their career if they perceive that their job is in danger.

During the *refreezing phase,* the individual displaces one set of views with another. Here, it is important to remember that even when they move away from their earlier views and biases, people can remain adrift and, consequently, discontent unless they internalize the new paradigm. This requires a commitment to, and an identification with, the change condition you wish to achieve. While training is an important tool for accomplishing this, it is equally important to demonstrate that the system values and rewards those who adopt this new paradigm.

Before leaving the subject of change management, let's examine a tool that may prove helpful in identifying where individuals stand in their willingness to adopt a particular change outlook. This tool, shown in Figure 11.3 and labeled the change-orientation matrix, is predicated on the reality that individuals typically pass through a series of stages when moving from the point of denial toward commitment to a particular change condition.

The vertical axis of the matrix highlights the fact that, in relation to any given stage, there is also a level at which the individual identifies with the change condition. Changes that individuals perceive as impacting them personally are naturally the most important to them, and the most difficult to effect. Those that impact something they value about their work team come next. Though changes that influence the organization

Change issue: Decentralizing purchasing operations

	Denial	Resistance	Indifference	Acceptance	Commitment
Individual	Dave Jones Sue Harmon	Barbara Nix Ed Herman		Al Volari	Stu Stewart Lucy Garcia
Team		Beth Black Bill Dent	Lucy Garcia		Uri Ambrose
Organization			Terry Carr	Sue Roman	

Figure 11.3 Change-orientation matrix.

may be important, such changes are subordinate, for most people, to those that impact self and team interests.

The change-orientation matrix is a practical tool for managing change. If we can determine where the various members of the process team stand *in relation to a specific change issue,* we are in a much better position to know what actions are needed to move them closer, as individuals, toward the commitment stage. The matrix can also assist in identifying the individuals who can serve as champions for change among their peers.

To complete the change-orientation matrix, listen carefully to, and then interpret, the comments individuals make when they are discussing a particular change issue. If, for instance, the change involves decentralizing purchasing, we might expect to hear such remarks as the following:

- The overhead on the process is going to increase substantially if we have to make our own purchasing arrangements. [This expression could be interpreted as *resistance* due to a perceived impact on the *team.*]

- I've never had a problem getting quick turnaround from the purchasing department. We've certainly managed to do business this way for as long as I've been with the company. [This is an expression of *denial* on the *personal* level.]

- If decentralizing the purchasing responsibilities is good for business, then I'm all for it. [This is an expression of *commitment* due to the perception of a positive *impact* on the *organization*.]

- This will definitely make it easier for us to purchase supplies, *but* it also puts the responsibility for purchase-order accuracy squarely on our shoulders. [This is an expression of *indifference* as it relates to an impact on the team. The use of the word *but*, however, can signal indifference or the existence of two or more levels of concern.]

- I've got mixed feelings about this change. I can see how it will cut out a layer of cost, but I also think it will increase my already overburdened workload. [This is an expression of *resistance* on the *personal* level and *acceptance* on the organizational level. This remark demonstrates that, for a given issue, an individual could be at different stages of acceptance.]

As Figure 11.3 suggests, a single matrix should be used for each change issue. The names of individuals can be placed in the cells that correspond to their attitude toward that particular issue, and a change-management response can be fashioned accordingly. This information should always be held in confidence and should never be used as an excuse for browbeating employees. If possible, base the assessment of where an individual stands on the opinion of at least two observers. In some cases, written surveys can also be constructed and administered for the purpose of collecting this information.

Impact on People and Jobs

There's no denying the fact that process reengineering can result in the loss of jobs. The inefficient use of human resources is a major cause for excessive costs and unnecessary delays. With their jobs potentially at stake, the process team simply cannot be objective about the need for major changes in the process nor about which changes are best from the customer's perspective.

The very real possibility that jobs may be on the line will certainly have an impact on the acceptance of any proposed changes. In fact, the resistance to change can turn into outright hostility when jobs are threatened. Furthermore, the term *process reengineering* will quickly acquire the reputation of being a euphemism for cutting heads, thus hampering future projects.

If the loss of jobs is inevitable, it is best to make this determination as soon as possible so that appropriate outplacement procedures can be followed. This measure will also help to alleviate the fears of the remaining team members that their own jobs may be on the line. The impact of layoffs should be weighed in advance and factored in the risk analysis, along with the possible need for a contingency plan.

Rather than jumping immediately to the conclusion that a cutback in the work force is necessary, it is wise to carefully consider other alternatives. For instance, could displaced workers be cross-trained to use their skills elsewhere in the organization? Looking further down the horizon, could improvements in process efficiency conceivably reduce costs, expand the company's client base, and, in turn, create the need for skilled workers who would otherwise be displaced?

Another very real possibility regarding people and jobs is that changes in job responsibilities often result from reengineering the process configuration. It is a common mistake to assume that process operators will simply know what to do when the new configuration is installed. Though we earlier mentioned the role of training in dealing with change, job-specific training may be necessary as well.

Related to training is the issue of certification. Even if proficiency certification has not been employed in the past, for some processes, you may need to identify the skills essential to each job position. If so, you would normally do this once the process changes have been installed and the process is stabilized.

This chapter has described a model for planning the transition. It takes into account the dimensions of change that often accompany process reengineering: physical, functional, organizational, and psychological. A well-thought-out transition plan not only serves as a control document for stepping through the implementation process, it also helps ensure early on that the reengineering team members (including senior management) all understand the game plan that will be followed in executing the transition.

Three of the more critical issues relating to implementation were also discussed in this chapter: phasing in the implementation, managing the human dimension of change, and paying careful attention to the impact that the outcome will have on people and jobs. The importance of these issues simply underscores the fact that the transition effort consists of considerably more than implementing functional changes in the business process.

CHAPTER 12

TRACKING AND MEASURING PROCESS PERFORMANCE

You simply have no way of knowing for sure if your process changes have produced the desired results unless you take meaningful performance measurements once the changes have been installed. For the sake of comparison, it is desirable to use the same metrics to evaluate the process in its reengineered configuration as you used in analyzing its original performance. Consequently, much of our earlier discussion regarding the metrics and techniques for analyzing the process in its current configuration is equally relevant to tracking and measuring its performance in its reengineered configuration. This chapter expands on that discussion and also introduces the concept of a *process performance measurement system*.

In assessing the performance of a process, whether before or after it has been reengineered, we are typically interested in two types of measurements: those that provide some indication of *overall* performance and those that tell us something about *what goes on within* the process. In either case, we must use care in selecting the parameters to be measured. Before any measurements are taken, the process reengineering team and upper management should agree on at least the following two issues.

- The parameters to be measured are valid and critical indicators of process performance.

- The methods that will be used to collect measurement data are valid and reliable.

Moreover, when contrasting the performance of a process in its reengineered configuration to that in its original configuration, the basis for comparison may be invalid if the *mission* of the process has been changed in any way, which is not an uncommon occurrence when a process has been radically overhauled. In other words, if the process now encompasses new responsibilities, or perhaps relinquishes others, those differences will have to be taken into account. This applies when comparing both *overall* performance metrics, such as process cycle time and process cost, and *in-process* measurements if the functional responsibilities of certain subprocesses no longer mirror each other.

If, for instance, a certain insurance company reengineers its claims-processing sequence such that one process handles only commercial claims while another handles only noncommercial claims, it may be meaningless to compare the overall performance of either new process with the original process that carried out both responsibilities. What may make sense in this situation is to compare the cost of the original process (denoted as C_o) with the cost of the two reengineered processes combined (denoted as C_{r1} and C_{r2}). If $C_{r1} + C_{r2} < C_o$, we at least know we are saving money. However, other factors may have to be taken into consideration, such as cycle time. If we factor in cycle time as well, the indication of success is somewhat more difficult to determine. If T_o is a measure of the cycle time of the process in its original configuration, and T_{r1} and T_{r2} represent the cycle time of the newly reengineered processes, then the following inequality conditions are possible:

Desirable outcome

$$C_o > C_{r1} + C_{r2}, \text{ and } T_o > T_{r1} \text{ or } T_{r2}$$
$$(\text{whichever is greater, } T_{r1} \text{ or } T_{r2})$$

Undesirable outcome

$$C_o < C_{r1} + C_{r2}, \text{ and } T_o < T_{r1} \text{ or } T_{r2}$$
$$(\text{whichever is greater, } T_{r1} \text{ or } T_{r2})$$

Questionable outcomes

$$C_o > C_{r1} + C_{r2}, \text{ and } T_o < T_{r1} \text{ or } T_{r2}$$
$$(\text{whichever is greater, } T_{r1} \text{ or } T_{r2})$$

$$C_o < C_{r1} + C_{r2}, \text{ and } T_o > T_{r1} \text{ or } T_{r2}$$
$$(\text{whichever is greater, } T_{r1} \text{ or } T_{r2})$$

If you have three parameters to consider—say, for instance, cycle time, process cost, and customer-satisfaction rating—you can have as many as eight possible inequality conditions. In this case, six of the eight outcomes offer questionable data as to whether the process has truly been improved. Clearly, then, process improvement is not always a black-and-white affair, even when accurate measurements are available. In such cases, interpretation also enters into the picture.

Notice that in general 2^n inequality conditions are possible, where n is the number of parameters we are interested in monitoring. Thus, a situation involving four parameters could result in 2^4 (or 16) possible inequality conditions. It's easy to see that the comparison of process configurations can be quite complex if a number of performance metrics are involved.

OVERALL PERFORMANCE MEASUREMENTS

As we have seen, two useful overall performance parameters are cycle time and process cost. There are a couple of points to keep in mind when discussing these two parameters. These points can be explained with the aid of Figure 12.1.

Notice that the process depicted by this top-level flow diagram has 12 major subprocesses or activities, some of which operate on parallel tracks (or run concurrently) with others. In calculating process cost, it would be necessary to determine the combined cost for all 12 of the subprocesses. By contrast, in calculating the cycle time, we are interested only in the longest duration path, since this particular path always defines the minimum time necessary to complete one full process cycle. Consequently, cycle time, by itself, does not give us a complete picture of the efficiency of the process unless, of course, cycle time is the only parameter we are interested in improving. Notice that one strategy for reducing the cycle time—if cycle time is of paramount importance—is to force more subprocesses to run concurrently, at least to the extent that this is possible. Those who are familiar with what is known as the critical path method (CPM) will recognize that the cycle time corresponds to the duration of the *critical path*.[1]

We may also be interested in monitoring overall performance parameters in addition to or instead of cycle time and/or process cost. Most of these are associated with one of three categories of metrics.

- Effectiveness ratings
- Efficiency ratings
- Adaptability ratings

Effectiveness ratings are directly tied to what the customer wants and values from the process. Such ratings typically tell us something about the accuracy, responsiveness, or reliability of the process. Here the primary issue is, How well does the process get the job done?

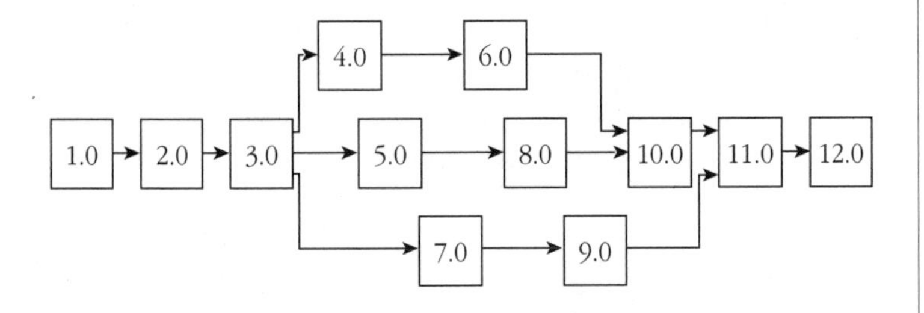

Figure 12.1 Top-level process-flow diagram.

Efficiency ratings are concerned primarily with judicious resource utilization. Efficiency metrics are of special interest to senior management, but they also are indirectly of interest to customers, since inefficiency contributes to process costs, which are ultimately passed on to them. Efficiency measurements may focus on costs, waste, rework, delays, resource utilization, or other forms of capacity utilization, to name a few.

Adaptability ratings tell us something about the versatility of the process. When adaptability is important, we would look for ways to measure the flexibility of the process in responding to an array of customer demands. Factors such as the ability to customize, make point decisions, and/or retain customer loyalty are directly, or indirectly, related to process adaptability.

PERFORMANCE RATIOS

We sometimes want to calculate a performance ratio based on a set of measurements. For the most part, performance ratios tell us something about the efficiency of the process from one vantage point or another. Here are some ratios that are often used in evaluating the performance of a process, but most especially in comparing the performance of two or more process configurations.

Cost to Cycle Time

This metric can be used for comparing process configurations having different cycle times and/or process costs, providing we keep in mind that it is desirable to reduce both factors: cost and cycle time. For instance, a cost-to-cycle-time ratio of 4:6 is better than a ratio of 5:6; however, a ratio of 4:6 is not better than a ratio of 4:5. If, for some reason, the costs decrease but the cycle time increases, it will be necessary to compare the *percentage decrease in cost* with the *percentage increase in cycle time,* or vice versa. In other words, if faced with the choice, we might be inclined to give up a little in terms of cost in order to gain a proportionately larger amount in cycle time. Frequently such trade-offs are not necessary, since improvements in cycle time are often accompanied by improvements in cost.

Real Value Added to Cycle Time

With this ratio, the idea is to determine what percentage of the overall cycle time actually comes from real value added activities (see Table 5.1). For a given process, the higher the ratio of RVA to cycle time, the better. Be aware that this ratio, when expressed as a decimal value, cannot tell us in absolute terms if one process configuration adds more value than another since a decimal ratio is unitless, but it does indicate whether one configuration is more efficient at doing this than the other.

Real Value Added to Process Cost

This ratio indicates the relative profitability of two or more processes, similar to traditional measures of return on investment. If RVA can be quantified in terms of dollars, then any ratio value greater than 1.0 would indicate a profitable process, at least before any overhead charges are allocated to the process.

IN-PROCESS PERFORMANCE MEASUREMENTS

Overall process performance measurements are important for assessing whether the process changes have accomplished what we had set out to accomplish in terms of end results, or business objectives. Nevertheless, more detailed measurements within the process are needed to indicate what is really going on. The precise nature of these measurements will depend on the intended application of the data. For instance, in analyzing the performance of an existing process, we may be interested in identifying and measuring sources of waste and inefficiency. Immediately following implementation of any changes, we may be interested in establishing a new baseline for the performance of each major subprocess, such as the work backlog at a particular work station. Then, once the reengineered process is stabilized, we will need measurements for tracking and controlling the process and for heading off problems before they arise. Getting to the level of detail needed to support each of these measurement objectives could require a breakout of the process flow to several levels of detail, as shown in Figure 12.2.

Notice the numbering convention used in Figure 12.2 to identify each process component in relation to its corresponding level of detail. In some cases, especially with highly complex processes, flow diagrams depicting four or five levels of detail

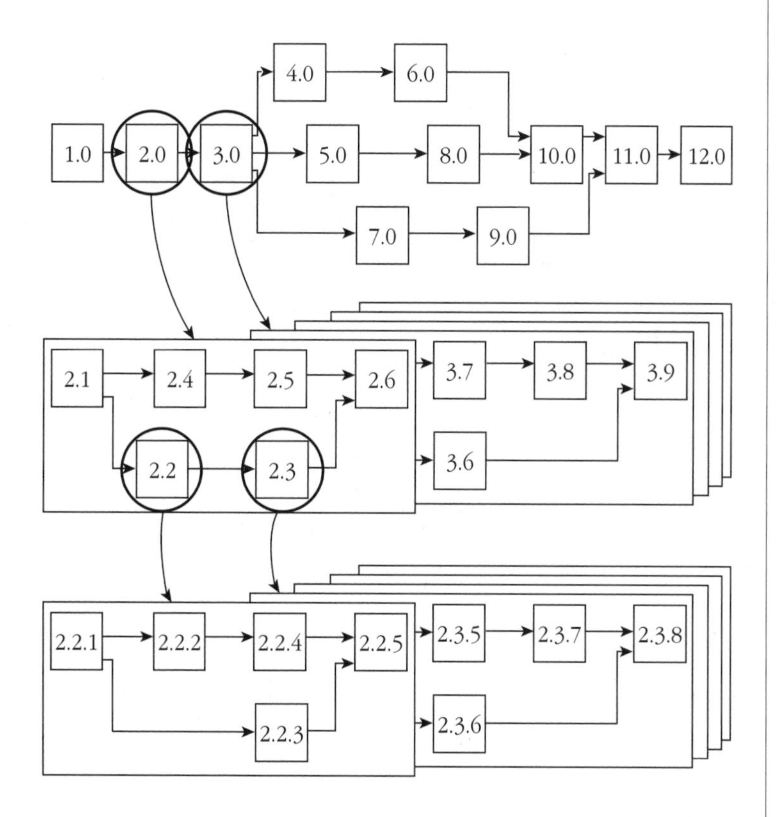

Figure 12.2 Process flow depicting three levels of detail.

may be required to identify specific points to monitor. Notice that the process flow lines can be broken down into greater levels of detail as well. For instance, at the top level, we may wish to indicate only the major connecting lines between the various subprocesses in order to depict the general flow of work. At the next level we may wish to add feedback lines, and, at the third level, any control lines or lines of authority.

Useful measurements within the process can be taken at any of a number of places. Here are some guidelines that may be helpful in identifying and selecting measurement points.

- Use quantitative data where possible.
- Identify measurement points that provide the most direct indication of high-potential problem areas. The FMECA technique described in chapter 10 may prove helpful in this regard.
- Pay special attention to interface points between subprocesses and functional departments. Subtle problems can develop at such points.
- Don't measure for measurement's sake alone. Be clear on why a particular measurement is needed and how best to obtain the data.
- Select measurement parameters that are relatively easy for process operators to obtain and understand.
- Use simple charting techniques to indicate process performance and to show any trend conditions. Post the charts for everyone to see.

PROCESS PERFORMANCE MEASUREMENT SYSTEM

Once the process changes have been implemented, a systematic program for monitoring and recording process performance should be maintained. By so doing, process variations can be observed over time, and problem areas can be more easily identified and corrected. Maintaining such records will also facilitate the reengineering process if this should become necessary in the future.

These are compelling reasons for establishing a process performance measurement system (PPMS) if such a system does not exist. Unfortunately, business processes have not kept up with manufacturing processes in regard to the entire issue of measuring process performance, let alone establishing a formal process performance measurement system. The basic concept for such a system is shown in Figure 12.3.

As this diagram shows, the PPMS is tied directly to the process improvement chain. Here is a brief description of the first three stages of the system.

Measurement

During this stage, measurement data are collected and recorded. The measurement process will conform to an agreed-upon methodology, such as statistical process control or position auditing. The measurement points should be selected according to their ability to clearly indicate where improvement opportunities lie and/or where problems are likely to surface ("windows of opportunity").

Evaluation

In this stage of the system, the data obtained from the measurement stage are evaluated in light of certain performance criteria. These may include internal criteria, benchmark standards, and/or customer specifications. Interpretation and problem-solving skills are of paramount importance in this stage.

Action

This stage is actually a decision point. Depending on the magnitude of variance between criteria and the measurement data, one of three courses of actions may be deemed appropriate: resolve the problem, make incremental improvements, or reengineer the process.

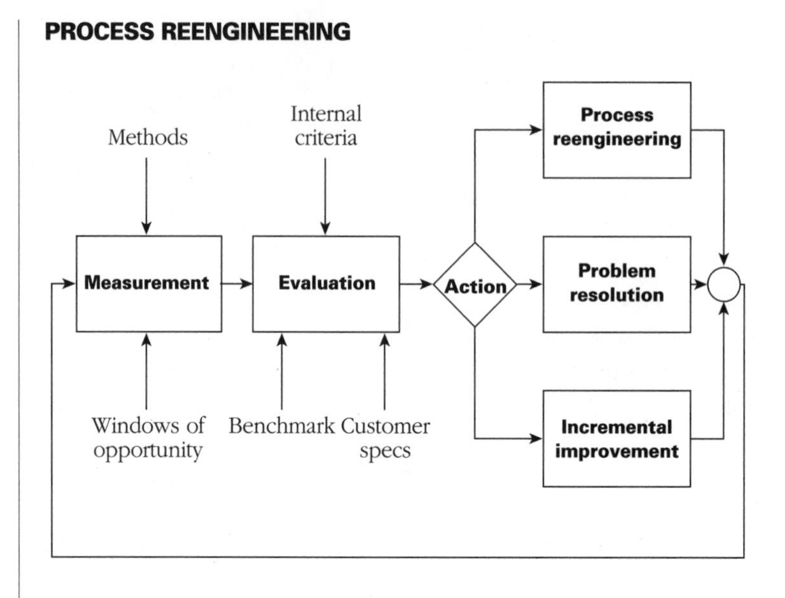

Figure 12.3 Process performance measuring system.

As Figure 12.3 also indicates, the PPMS is a cybernetic process, since the output continuously feeds back into the input, creating a self-regulating cycle. Even when the margin of difference between the expected performance and the actual performance of the process is below the threshold for taking corrective or preventive action, it is still necessary to measure and evaluate the performance of the process periodically.

As with most initiatives, the difficulty comes in implementing and maintaining the PPMS. Upper management must clearly support such a system if it is to be sustained and used appropriately.

This chapter has focused on tracking and measuring process performance, primarily with the intent of comparing performance before and after radical changes have been made. Tracking, and especially measuring, the performance of any business process can be a painful undertaking to say the least. There is little wonder why this is painful—with the myriad of business processes that have slowly evolved and lumbered along over the years revealing process efficiencies on the order of 1 or 2 percent. While the process reengineering movement did not invent the need for tracking and measuring process performance, it has certainly added impetus to the cause.

REFERENCES

1. The Critical Path Method (CPM) is a widely used project management tool that has been in use since the early 1960s. Further information on CPM can be found in most any text on project management, as for instance, Jack R. Meredith and Samuel J. Mantel, *Project Management: A Managerial Approach,* 2d ed. (New York: John Wiley & Sons, 1989).

CHAPTER 13

ADVANCED TOOLS AND TECHNIQUES

The basic concepts of process reengineering can be adapted to virtually every identifiable business process, regardless of its size or complexity. Even a single individual, acting under the authority of a department head, can successfully reengineer an intradepartmental process, assuming he or she has the organizational skills necessary to coordinate the project. Nevertheless, large, complex processes present some special challenges, because they include a multitude of details and partly because the resistance to change increases with the number of people who will be affected. Process optimization is also more difficult with such processes.

Fortunately, a number of advanced tools and techniques may help the project team with at least certain aspects of the project. The purpose of this chapter is to introduce some of the more powerful tools and techniques that are available. We will limit our discussion to a brief description of each tool or technique and examine its potential contribution to the process reengineering effort.

The reader is encouraged to seek additional information from other sources, such as an independent consultant and/or product vendor, if it appears a particular tool or technique is suited to a specific application. Also, product vendors often have demonstration packages available that will allow the potential user to preview a product before making a purchase commitment. Since there is often a lengthy learning curve and considerable cost associated with any software tool, it is prudent to seek a testimonial from at least two end users before making a capital purchase. Beware of product vendors who are unresponsive to your request for information or who use doublespeak and esoteric language to describe their product. Moreover, the use of discipline-specific language could reveal a bias that favors a certain reengineering solution-orientation that may, or may not, be best for the process in question. Furthermore, be sure you understand the hardware requirements for a particular software product before making a purchase. The garden-variety personal computer often lacks the speed, memory (both RAM and hard disk capacity), and/or sophistication (for example, client-server network) to support many vendor products or to take full advantage of their stated capabilities.

Some tools and techniques described in this chapter have had their genesis in the so-called quality movement. Others are associated with disciplines such as systems engineering, information systems, and project management. This overlap simply highlights the fact that process reengineering is, itself, an interdisciplinary endeavor that often brings together a team of individuals having an array of skills, abilities, and job responsibilities.

ADVANCED TOOLS

A mechanic charged with overhauling an engine would not think of starting such a job without having access to the right tools. This analogy is relevant to the project team charged with overhauling a business process. The flip side, of course, is that a relatively small job does not require a sophisticated set of tools with more bells and whistles than the job demands.

Before embarking on a reengineering project, you should learn what tools are available and how they can support the project. While no tool can substitute for human reasoning—the kind of reasoning that reengineering effort demands—certain tools can help with one or more of the following tasks.

- Recording and structuring data
- Supporting decision making (e.g., decision support systems)
- Diagramming process flows
- Isolating flow paths according to user-specified conditions
- Calculating process metrics
- Optimizing flow paths
- Modeling and simulating process layouts
- Performing dynamic what-if analyses on flow paths
- Pinpointing bottlenecks in the process
- Generating reports and plotting process-performance statistics
- Planning and tracking the reengineering project
- Diagnosing team behavior, values, resistance to change, and so on

With this list of capabilities in mind, let's look at some of the more advanced tools available to support the process reengineering effort.

Automated Project-Management Tools

Numerous software tools fall within this category. On the low end are products that do little more than simplify the creation of

Gantt charts (which may be all that is needed to support a relatively small project). Products on the high end can assist in organizing data, creating critical path method (CPM) charts, tracking costs and schedules, planning resource requirements, and dynamically testing what-if conditions, to name a few. Some packages also have features to assist in coordinating multiple projects.

While these tools are used primarily to support project management, certain features may be adapted to analyzing and reengineering the business process. For instance, in some cases it is possible to depict process flow using a CPM chart. With this chart, the process metrics can be calculated directly, and the process cycle time (which equates to the critical path) can be easily determined and instantaneously reestablished as changes in the flow sequence are tested. There are, however, some serious limitations in using CPM charts to model process flow. For one, CPM charts cannot depict more than a single link between any two activities. For another, the rules of convention do not allow for the existence of feedback loops. Nevertheless, with a little imagination, it may be possible to work around these limitations.

From time to time, an organization known as the Project Management Institute does a comparative study of some of the project management software tools that are available.[1] Their analysis may prove helpful in exploring software options.

CASE Tools. Computer-assisted software engineering (CASE) tools were introduced during the 1980s in response to the concern that software projects were becoming exceedingly complex and difficult to manage. CASE tools held the promise that the dozens of hours required to produce even a few lines of fully executable code could be reduced to a mere fraction of that time. For projects having tens of thousands of lines of code, this promise was hard to ignore, though compromises, such as the inherent reduction in program-execution efficiency, were found to be more than some applications could tolerate.

Since their introduction, CASE tools have become increasingly sophisticated. Over time, certain products have essentially become integrated software design and project management tools. For instance, some CASE tools can produce process flowcharts, create data dictionaries (that allow flow lines to be labeled and later referenced), and design prototypes of computer screens as users would view them. Other CASE tools can support many of the same tasks supported by project management software tools, such as planning and tracking schedules and budgets. CASE tools have at least one drawback: they typically lock the user into a particular flowchart format. This format may be excellent for modeling data flow paths but may not necessarily be the best for depicting document or product flow paths. The format can also be confusing to nonspecialists who need to be able to read and understand the process flowcharts. Unless your organization is currently using a particular CASE tool and this tool has the features you need, you may be able to find better alternatives for supporting the various process reengineering projects.

Workflow Systems. Business processes often consist of a number of well-defined steps that are essentially repeated with every process cycle. These steps are sometimes referred to as *production-based business processes.* To some degree, such processes are found in every organization, often in the form of billing, inventory control, purchasing, and so on. Indeed, production-based processes are the mainstay of service organizations such as lending institutions, securities brokers, and insurance companies. They also play a significant role in hospitals, utility companies, airlines, and telecommunications companies, to name a few.

A characteristic attribute of production-based processes is that they almost always involve the processing of information that, in some format, is routed from point to point within the process. The prevalence of such processes, and the inherent inefficiency associated with shuffling documents, forms, memoranda, and the like from one workstation to the next, has given rise to computer-based systems, known as *workflow systems.* Workflow systems come with a variety of features, but one thing they have in common is a means of mapping the workflow and automatically routing the work product between the various points within the process as it has been previously defined by the map. When work is completed at one station, the work product is automatically routed to the next station.

First generation workflow systems were rigid in the sense that extensive software coding was necessary in order to define and redefine flow paths. Today, most commercially available workflow systems allow a process administrator, as an end user, to create the maps and make changes in the process flow paths as needed without extensive programming.

From a process reengineering standpoint, the most important feature of workflow systems is that they allow the process in question to be easily modeled and tested before being implemented. The actual workflow routing capabilities of such systems are only beneficial once the process configuration has been selected and deployed.

Imaging Technology. As stated earlier, production-based business processes typically involve the processing of information in some form or another. Examples might include insurance claims, hospital admissions forms, engineering change documents, business proposals, and purchase orders.

Paper has traditionally been used as the medium for transporting such information from point to point within the process. But with the proliferation of copying machines, desktop printers, and fax machines—combined with the hoard of forms and other documents that now accompany most business transactions—organizations have literally become swamped in paper. Furthermore, paper documents have the nasty habit of getting misplaced, misrouted, and misfiled, not to mention getting buried for weeks in someone's in basket.

As a result of the numerous problems associated with paper forms and documents, some companies are beginning to migrate toward the use of electronic imaging systems. For instance, rather than passing a loan application form from one person to the next within a certain processing chain, a lending institution might instead replicate the form on a computer screen. Once the form, and the information recorded there, are available in electronic format, anyone who has access to the computer network can easily call up the form together with its accompanying information. Even when paper documents are unavoidable, such as the original loan application form completed by the client, the form and its contents can be read into the imaging system via a scanner and then stored or routed electronically—perhaps using a workflow system to control the routing.

While many reengineering successes have resulted from the marriage of imaging technology and workflow systems, this is a good place to remind ourselves that process reengineering should begin at a more fundamental level. Again, the decision to seek a technology-based solution should be made only after the reengineered process configuration has clearly been defined by what the customer wants and values. Automating a poorly designed process will do little to improve productivity.

In light of the discussion on workflow systems, it is helpful to examine specific features and capabilities of an actual tool. The descriptions that follow are neither an endorsement for, nor an evaluation of, a particular product.

***FloWare*™.** FloWare™ is a commercially available workflow system, developed by Recognition International, that runs on a client-server computer network. This system consists of the five following major components.

- **MapBuilder.** This component is used to graphically construct the process flowcharts, or *maps,* as they are referred to by this and other workflow product developers. One feature of MapBuilder is that it allows information, documents, or other work products to be dynamically represented as they flow through the process. Moreover, information flow paths can be easily moved, eliminated, split, or joined. Process activities, represented by user-designed icons, can also be added, deleted, or repositioned with ease. Finally, submaps can be added at a given level of representation to allow for expansion of detail where necessary.

- **Trailer.** This component maintains the records and statistics on each piece of information (or other work product) as it passes through the various activities within the flow path. These records—which may include a date and time stamp, a user identification, and/or a description of how the information is used or modified by a certain activity—are dynamically logged into a database for later access. With the process represented in this manner, it is possible to isolate work flows from the activities that support them. This representation can be especially helpful in analyzing and modeling situations where a single activity, entity, or department is responsible for supporting multiple processes.

- **Status Monitor.** This component allows the process administrator, or other members of the business process team, to check the status of work in process, determine how many users are logged into a particular activity, and/or check the priority of the activity. The Status Monitor is especially helpful in identifying potential bottlenecks, which is important from a process-improvement standpoint.

- **Exerciser.** Used in conjunction with the MapBuilder, the Exerciser allows the reengineered process to be modeled and simulated before being implemented. This feature allows bottlenecks and other inefficiencies to be identified in advance. Parameters, such as the volume of work and the timing of activities, can then be adjusted on a what-if basis to optimize the flow of the process.

- **User Administration.** This component defines the activities a certain process worker is allowed to perform. It also allows the process administrator to add, delete, or update user account information in the FloWare database.

Business Design Facility™. A number of software-based tools specifically designed to support process reengineering applications have begun to appear on the market. One such tool is Business Design Facility™ from Texas Instruments, which was commercially available as Release 1 in May 1993.

When available in its full implementation, the Business Design Facility (BDF™) will consist of three major components: Data Capture, Business Modelling, and Analysis. Here is a brief description of what each component does.

- **Data Capture.** As its name implies, this component supports the early planning stages of the reengineering project by allowing users to record and structure data that originate in meetings and planning sessions. Traditionally, ideas generated during such meetings are recorded using the ubiquitous yellow notepads, and then strung together on a whiteboard, perhaps as a means of depicting the subprocesses and their interrelationships. The adhesive sheets are moved from place to place on the whiteboard as the subprocesses and their connecting links are discussed and brought into sharper focus. In essence, the Data Capture component automates most of this task. When used with a video projection system, this component can serve as a substitute for, or adjunct to, the manual method of capturing and organizing data in a group setting. (Note: The Data Capture component was not incorporated into Release 1.)

- **Business Modelling.** The Business Modelling component is the real workhorse of the system and perhaps the most valuable from a process reengineering standpoint. This component allows the business process to be represented in diagrammatic form. The process activities can be displayed in a hierarchical format, similar to an organizational hierarchy, or in the more traditional process-flow format. Activities and their connecting links can easily be added to, removed from, or shifted within the model, saving much of the labor that would otherwise go into creating and modifying the process flowcharts.

The activities and their connecting links are labeled and assigned to a particular flowchart level. Then, the flowchart can be expanded or collapsed at will, depending on the level of detail the user wishes to view. Activities that appear at different levels of detail can also be *encapsulated* so that the view of one or more parts of the process can be magnified as necessary. The Business Modelling component also allows user-defined process metrics to be assigned to the various activities. These metrics can then be linked to an electronic spreadsheet (through a dynamic data exchange, or DDE, interface), where statistical calculations can be performed as the user may require.

- **Analysis.** Technically speaking, the Analysis component is external to the BDF system. As noted above, however, the BDF does provide the DDE interface links necessary to transport data "on the fly" to electronic spreadsheets, word processing systems, or other commercially available software packages that can communicate via the DDE protocol. A component called *Open Interface* is also available to allow for data links with other commercial products such as CASE tools and activity-based costing packages. Texas Instruments has its own CASE tool, which can interface directly with the BDF system without passing through the Open Interface.

Change-Management Tools

We have repeatedly stressed that the human dimension can represent the greatest challenge to the process reengineering team. Given the relative importance of this issue, it's ironic that, by contrast, so little thought is given to the use of tools that can support the change-management aspects of the project.

The change-management process could benefit a great deal from insight into such factors as individuals' attitudes toward change, their views on how much they can actually influence change, or their ability to work as a team to support the process. A broad array of such tools can be adapted to provide this, and other, insight into the attitudes, perceptions, and reasoning styles of the process team.

One such tool is the Problem Solving Style Assessment™.[2] This tool profiles the cognitive style of an individual according to each of five thinking and problem-solving modalities: analyst, realist, pragmatist, idealist, and synthesist. Potential strengths and weaknesses are identified, and team compatibility factors are described. This information may help you plan the process changes in relation to each individual or match up process team members to achieve compatible (which is not to say similar) problem-solving and reasoning styles.

Another tool, the Fundamental Problem Attribution Scale™,[3] assesses the degree to which individuals perceive that they are in control of their environment. This instrument also measures the individual's perception of whether he or she is an instigator or a passive observer in relation to changing or maintaining the status quo. In a process reengineering context, this information may be helpful, for instance, in identifying those individuals who will most likely resist change because they fear they are not, but should be, in control of their circumstances.

ADVANCED TECHNIQUES

Now that we have surveyed certain advanced process reengineering tools, let's examine several advanced techniques as well. The techniques described below are most relevant to large, complex processes. In some cases, adaptations may be possible to benefit applications on a smaller scale. A brief description of each technique is provided along with a statement regarding the technique's potential relevance to process reengineering applications.

Multiattribute Utility Analysis

If optimizing a certain process simply requires us to minimize cycle time at whatever cost, then the selection of a process configuration would be relatively easy. The same goes for minimizing the cost of the process. But, when two or more criteria have to be considered, such as minimizing cycle time and process cost, it may be impossible to accomplish both objectives fully with a single process configuration.

The point is this: selecting an optimum process configuration can become extremely complex when trade-offs in decision variables must be considered. Fortunately several techniques can be employed in such cases. Multiattribute utility analysis is one such technique.

Multiattribute utility analysis allows a team to select an alternative that optimizes the trade-offs between two or more decision variables. Here is a brief description of the technique:

1. Each decision variable is identified and then assigned a relative weight (which we will call W).

2. The decision makers then assign three values to each decision variable: a desired value, an unacceptable value, and an indifference value. For instance, for total process cost, we may establish \$2000 as the desired value, \$4000 as the unacceptable value, and \$2600 as the indifference value. (The *indifference value* represents the point at which a particular decision variable is judged to be neither good nor bad.)

3. A graph, such as that shown in Figure 13.1, is then plotted for *each* decision variable, using the three values defined for each variable. The horizontal axis displays the range of values the variable can assume (\$2000 to \$4000 in our example). A scale ranging from 0 to 1.0 is then created on the vertical axis, where zero corresponds to the unacceptable value and 1.0 corresponds to the desired value. The indifference value sits at the 0.5 point on the vertical axis.

4. From the vertical axis of each graph, a *utility value* (which we will call U) is determined. The utility value represents the actual value that a particular decision variable assumes. (In our example, if we

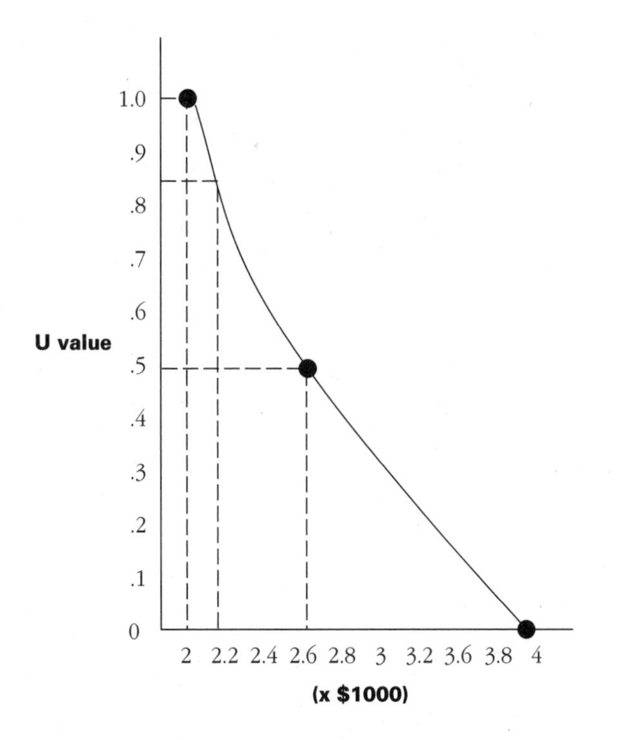

Figure 13.1 Utility curve for process cost.

were to determine that the total cost for a certain process configuration was $2200, then the corresponding utility value would be 0.85, as determined from the graph.)

5. A *weighted utility value* is then computed by multiplying $U \times W$ for each variable. (If the weight assigned to this particular variable is 5, then the weighted utility value, $U \times W$, would be 5 x 0.85, or 4.25.)

6. Next, the weighted utility values for all decision variables of a given process configuration are added together to get a composite index C for that particular configuration.

7. Steps 4 through 6 are repeated for the other process configurations. The process configuration with the highest composite index is selected as the optimum alternative.

Goal Programming

Goal programming is a quantitative technique that can help in making decisions regarding the optimum allocation of resources.[5] This technique also allows the user to consider constraints and multiple objectives. Since goal programming is concerned with what could be rather than what is, it is better classified as a synthesis technique than as an analysis technique.

This technique might be used to allocate human, physical, and/or financial resources to the various process activities when a certain attribute of the process can be regulated or adjusted. If, for instance, we have the ability to control the number of people assigned to each activity *and* we have some way of determining each activity's contribution to the various objectives, then a goal programming model could be constructed to determine the optimum allocation of people to each activity.

Goal programming involves the use of matrix algebra. Computer programs capable of executing the necessary calculations and matrix manipulations are commercially available. Although the decision makers who design the model do not need to understand matrix algebra or the algorithms employed in the computer program, they do need to set priorities among the objectives and judge the contribution of each variable to the performance of a particular activity.

Design of Experiments

One drawback with multiattribute utility analysis and goal programming is that neither technique can take into account the possible interactions between process variables. When, for example, we have the ability to regulate the number of people assigned to the process as well as the size of the work space, it is clear that the addition of people will be influenced by the size of the work space. Efficiency will eventually start to diminish as people become packed into a tight work space. It is in situations such as this, where process variables may interact, that the design of experiments (DOE) technique is best suited.

Certain DOE techniques also allow the process designers to use statistical shortcuts to set up experiments and test how the process might behave if two or more process variables are involved *and* if each variable can assume two or more values. For instance, if there are four variables, and each can assume three values, it is possible to identify 3^4, or 81, combinations of variables and levels that would potentially have to be tested. The shortcut procedure allows us to accomplish essentially the same results by setting up and running only nine experiments.

Design of experiments is a sophisticated technique that, even with the shortcut methods, involves considerable time and effort to run. But it may be possible, depending on the variables involved, to manipulate the variables under simulated conditions (as with a computer model) rather than in actual experiments.

The reengineering team should seek the help of an internal or external consultant who thoroughly understands this technique if it appears that DOE may benefit the reengineering initiative. In any case, the team need not understand the theoretical underpinnings of DOE to benefit from its results.

This chapter has surveyed a number of advanced tools and techniques that can be brought to bear on the reengineering project. The primary intent has been to enlighten rather than expound, which is in keeping with the fact that the reengineering team should stay focused on its primary task: revamping the process. Nonetheless, the team, and especially whoever fulfills the role of reengineering consultant, should be aware of the tools and techniques that are available to support the project and of what they can contribute to the project's success.

REFERENCES

1. Project Management Institute, PO Box 43, Drexel Hill, PA 19026.

2. The Problem Solving Style Assessment™ is a product of Roberts & Roberts, Plano, Texas.

3. The Fundamental Problem Attribution Scale™ is a product of Roberts & Roberts, Plano, Texas.

4. Thomas A. Payne, *Quantitative Techniques for Management: A Practical Approach* (Reston, Va.: Reston Publishing, 1982).

5. James P. Ignizio, *Goal Programming and Extensions* (Lexington, Mass.: Lexington Books, 1976).

6. William J. Diamond, *Practical Experiment Designs for Engineers and Scientists* (Bellmont, Ca.: Wadsworth, 1981).

CHAPTER 14
CASE PROFILES

Now that we have seen what goes into a process reengineering project, let's wrap things up by profiling several cases. For proprietary reasons, companies are understandably reluctant to specify what they have done in the way of process reengineering and how they have done it. They are even more reluctant to publicly acknowledge that they harbored waste and inefficiency in their business processes for many years, despite the fact that conditions may have dramatically improved. As a result, only sparse information is available in some cases. Nevertheless, if we glean what we can from these cases, we can pool this information in the form of lessons learned.

In studying these profiles, you may notice similarities as well as dissimilarities in philosophies, approaches, and techniques. The dissimilarities simply highlight the fact that circumstances unique to each organization dictate the course of a reengineering project. While some projects described here were completed before the label *process reengineering* became popularized, they embody enough features of process reengineering to warrant labeling them as such.

PROCESS REENGINEERING

FORD MOTOR COMPANY

Inspired by the efficiency of the accounts payable process at Mazda—which was supported by five people in contrast to the five hundred people that Ford employed for this purpose in North America alone—Ford managers began a serious examination of how to radically improve their own accounts payable process.[1] Analyzing the existing system, they discovered that a major cause of inefficiency was the frequent mismatching of purchase orders issued by the purchasing department, receiving documents sent to accounting when goods arrived, and invoices sent by the vendors. Too much time was being spent trying to reconcile these mismatches. In the meantime, payment to the vendor would be held up while a clerk investigated the discrepancy. And even more paperwork would be generated in the process of trying to rectify the problem.

Ford's reengineering solution resulted in invoiceless processing. In its new configuration, the accounts payable process was set up to link purchasing, receiving, and accounting to a common database so that matching could occur electronically rather than manually. Furthermore, the *procedures* were changed so that only three items would need to be matched, versus fourteen under the old system. The requirement for vendor invoices was also eliminated. Now, when an order arrives, the receiving department immediately logs this information into the shared database. Accounting issues a check to the vendor shortly thereafter, relying only on the information logged in the database by purchasing and receiving.

What benefits was Ford able to realize? In the portions of the company where the reengineered configuration has been installed, Ford has reduced its work force by 75 percent. Furthermore, financial records are now more accurate due to the elimination of matching problems.

WESTINGHOUSE ELECTRIC

In the following two cases, Westinghouse Electric reengineered internal process, first streamlining and then defining and installing technology requirements to implement the reengineered process. Even though little more than the outcome has been reported,[2] the results are certainly convincing.

The first case involved the customer-ordering process in one of the company's businesses. By restructuring and then automating the order-entry procedures, the time needed to process an electrical parts order was reduced from 28 hours to just 10 minutes—an amazing 99.4 percent reduction in cycle time! Along with this reduction, the cost per order was cut by two-thirds.

In the second case, the cycle time required to reload fuel assemblies in nuclear power plants was reduced from three years to eighteen months. This reduction resulted in a 25 percent reduction in costs. The reduction in cycle time also allowed the engineering department to increase its workload by 40 percent, with only a 10 percent increase in head count.

BELL ATLANTIC

When Bell Atlantic examined the cycle time involved in filling an order for telephone lines from a long-distance carrier, it found that the order passed through 28 hands before being filled.[3] This was attributed to the company's function-based structure, which carved work in process into discrete tasks that were then allocated among departments—some being geographically dispersed. Each department would dutifully perform its tasks and eventually pass the work in process on to the next department.

Bell Atlantic made sweeping changes in its order-placement process by eliminating tasks of lesser importance. It also assigned teams to study the process by following a single order as it passed from one end to the other.

When the decision was made to pursue radical improvements, Bell Atlantic established two reengineering teams, each having a different charter. One team, called the core team, was responsible for generating ideas and redesigning the process, while the other team was responsible for testing and refining the ideas put forth by the core team.[4]

The results? The cycle time for the order-placement process, which previously took from 15 to 25 days, was reduced to a matter of hours, with a future goal of diminishing this to a matter of minutes. Furthermore, because it is now more responsive to the customers, the company has begun winning back market share from its competitors—the alternate-access carriers.[5]

JOHNSON FILTRATION SYSTEMS

Johnson Filtration Systems makes water- and oil-well screens and filters. In 1990, the company began a total quality management (TQM) initiative and sought greater employee involvement.[6]

An extensive training program was put in place to familiarize everyone with the concepts and to convert the skeptics and secure commitments from the fence sitters. A crossfunctional team, consisting of 18 carefully selected members, was then established to identify problem areas and recommend solutions. However, employees achieved breakthrough levels of improvement only when the task force was divided into smaller teams that focused on three specific issues: paperwork, materials, and manufacturing. The smaller teams met once a week over a 13-week period and pored over process flowcharts while forging a new process design.

The results were respectable, if not astounding. Cycle time went from 28 days down to 3 days, a 9.3 to 1 improvement. But these 3 days later built up to 5 days, due to what one official characterized as "communications problems." Still, the number of workers handling the product went from 28 down to 7, or a 4 to 1 improvement.

XEROX

In 1992, Xerox initiated what its CEO labeled "reengineering the enterprise."[7] In essence, this amounted to a major restructuring of the organization to become better focused on what it considers its primary business: document systems.

Even though this reengineering effort was instigated from the top, such is not necessarily the case with other reengineering projects at Xerox.

> *Much of the day-to-day leadership of Xerox's reengineering projects originates further down the corporate ladder. Each of three Xerox reengineering projects—time-to-market, proposal order to collection, and integrated supply chain—is being led by a project team in partnership with IS (Information Systems) staffers.*[8]

Since its first business reengineering project in 1986, Xerox has learned some difficult lessons about what does and doesn't work. Here are some of the more important discoveries that Xerox made along the way.

- Reengineering projects are not likely to succeed if the information systems group is running the project.
- CASE tools are incomplete as they pertain to process reengineering projects. These tools have you "studying information links, not processes."
- Process planning must come before technology planning.

Perhaps these points mask an important observation regarding Xerox itself. Namely, in many respects, Xerox is exhibiting the same ability to take risks that it did in the late 1950s, when market research failed to substantiate the need for a business copying machine. Obviously, the formula for success, when it comes to anything, including process reengineering, goes beyond a list of dos and don'ts. The culture of the organization is a key factor as well.

TEN LESSONS LEARNED

The cases profiled in this chapter and those described elsewhere collectively give us a picture of the success and failure factors associated with the process reengineering approach. We list the following as ten lessons learned, without elaboration and in no particular order, primarily for the benefit of those who are contemplating a reengineering project but have no history of reengineering within their organization to draw on.

1. **Never forget that reengineering begins and ends with the customer.** Get to know what your customer really wants and values. From this, establish the critical success factors you will use to gauge the before and after performance of your business process. Don't just *listen* to the voice of the customers. Discover ways to delight them. Such thinking is necessary to keep your organization a step ahead of the competition.

2. **Pay particular attention to the human dynamics of change.** The success of the project depends on people's attitudes and willingness to buy into the process as well as the outcome. Be prepared to deal with the human aspects of the project. They will be harder to manage than the technical aspects—and are more pivotal to the success of the project as well.

3. **Get senior management involved in the reengineering project from beginning to end.** Choose an executive sponsor to shepherd the project. Find a person who has a reputation for getting things done yet knows how to engage and motivate the project team. Give this person the final word on choosing the project team members. Also ensure that he or she has the authority necessary to overcome interdepartmental barriers.

4. **Prepare people for change.** Don't leave them in the dark. Don't assume that they will be naturally resistant

to change, yet keep in mind that people may have to go through a grieving period before buying into the process or the outcome of change. It is better that such grieving occurs early—not when you need people to be on board with the project.

5. **Walk through your existing business processes.** Observe how work gets handled along the way. Pay attention to, and carefully document, sources of waste and inefficiency, points in the process where work gets recycled, aspects of the process that do and do not add value from the customer's perspective. Notice how the process workers handle problems, and measure how the process performs in relation to the critical success factors. Pay particular attention to the interface points between departments, functional entities, and individuals. Use this information to create an as-is map of the process, but don't fall victim to the analysis paralysis trap—especially since you're looking for innovative ways of doing things.

6. **Pick the best available people to serve on the project team.** Consider not only the need to represent the viewpoints of the various functional entities that support the process, but also the mix of skills, abilities, and attributes that the team members should possess. Replace any team member—including the team leader or even the executive sponsor—if, and as soon as, it becomes apparent that that individual, for whatever reason, is an impediment to progress.

7. **Think boldly and go for a win.** Go beyond the limits of what seems possible for satisfying the customer. Start with this stretch goal and work backward to define the optimum process configuration for achieving the goal. Leverage your success, but don't penalize yourself for thinking boldly and possibly underachieving your initial goal.

8. **Keep technology and automation in perspective.** First determine how the process should be structured from a functional standpoint; then examine the technology alternatives for making this structure even more efficient. Don't allow "sunk costs" in technology or automation to influence your perspective on what is best for now and the future. Ensure that the reengineering project is owned by operations specialists rather than information specialists.

9. **Think in process terms.** Ensure that the organizational infrastructure exists to support your core business processes—not vice versa. Appoint process managers to oversee these processes from end to end. Give broader responsibilities to the individual process workers, and empower them to make decisions, solve problems, and communicate directly with the customer. Train and certify the process workers to support the process in its reengineered configuration.

10. **Be persistent.** Don't be discouraged by can't-do thinkers or those who parrot the line, "We've been down this road before." Set intermediate goals, and celebrate the success of achieving these goals. Carry out your mission as though the very survival of the organization depends on its success—for indeed it may!

As we have repeatedly demonstrated, companies that have applied the process reengineering approach to their business processes have realized quantum degrees of improvement in cycle time and cost reduction. But these improvements could be only the beginning, since process reengineering fosters the kind of innovative thinking that gave rise to most corporations in the first place—especially before they built walls of isolation that fragment their business processes.

Is process reengineering a passing management fad, as some would contend? The answer is yes if reengineering is given half-hearted commitment. But the answer is no—definitely not—if we accept that fact that dramatic improvements in our business processes are long overdue and that the need to occasionally challenge our basic assumptions about the way we do business is not an issue of what is or isn't fashionable. The real question is not whether reengineering is a passing management fad, but rather, What took us so long to realize that process reengineering represents a smart way to run a business?

REFERENCES

1. Michael Hammer, "Reengineering Work: Don't Automate, Obliterate," *Harvard Business Review* 68 (July-Aug. 1990): 104.

2. George C. Dorman "Go with the Flow—Measuring Information Worker Quality," in *Total Quality Performance,* ed. Lawrence Schein and Melissa A. Berman (New York: The Conference Board Research Report no. 909, 1988).

3. John A. Byrne, "Paradigms for Postmodern Managers," *Business Week,* 1992 Reinventing America Issue, 62.

4. Michael Hammer and James Champy, *Reengineering the Corporation: A Manifesto for Business Revolution* (New York: HarperCollins, 1993).

5. Thomas A. Stewart, "Reengineering: The Hot New Managing Tool," *Fortune* 128 (23 Aug. 1993): 41.

6. Marc Hequet, "Quality Concerns Come Through Unfiltered at Johnson Filtration," *The Quality Imperative* (Minneapolis, Minn.: Lakewood Publications, 1992).

7. Peter Krass, "A Delicate Balance," *Information Week* (supplement) (5 May 1992): 26.

8. Ibid., p. 28.

GLOSSARY

Business process reengineering Same as process reengineering: the application specifically refers to business processes rather than manufacturing processes.

Business value added activity Any business activity or subprocess that is essential for conducting business but that does not add value from the customer's perspective.

CASE Acronym for computer-assisted software engineering.

Change management A proactive approach to managing the human dimension of change; considers the incentives and disincentives that influence the acceptance of and commitment to a particular change paradigm.

Change-orientation matrix A matrix used to identify where individuals are positioned in relation to the acceptance or rejection of a given change.

Critical path method A network-diagramming technique depicting the precedence relationships of a set of project tasks. The diagram uses arrows to connect the tasks. The critical path refers to the longest time-duration path. Any delays in the critical path will delay the overall project.

Cycle time The amount of time it takes for a certain process to complete a job.

Design of experiments A technique for identifying the optimum process-performance conditions in situations where two or more variables that are related to that process can be manipulated.

Executive sponsor An executive who has been appointed to serve as a member of the process reengineering project team. This individual provides a critical link to upper management while helping the project stay on track and not become bogged down by interdepartmental barriers.

Failure analysis A systematic approach to identifying, anticipating, and heading off potential problem areas and weaknesses in a certain process configuration before it is deployed.

Infrastructure Includes the facilities, technology, systems, and equipment that are needed to support the business process but that are not technically a part of the process itself.

No value added (NVA) activity Any business activity or subprocess that the customer does not need or is not willing to pay for, or that does not fulfill a business requirement.

Process capability Describes what a process, as it is designed and normally operated, is capable of producing, in contrast to what is desired. A process could be in control but, due to its design, incapable of staying within the desired tolerance limits.

Process experts Individuals who have specialized knowledge of a certain process. A process expert appointed to serve on the reengineering project should also have broad knowledge of the overall process.

Process flowchart A diagram that depicts the flow of information, paper, or other work products, from one subprocess or activity to the next throughout the process. Also referred to as a *process map*.

Process owner The individual who has end-to-end responsibility for a certain business or manufacturing process.

Process performance measurement system (PPMS) A systematic approach to monitoring the performance of a process; specifies what to measure, the bases for interpreting these measurements, and perhaps, the thresholds that trigger a certain response.

Process performance metrics The indexes that are measured in order to characterize the before and after performance of the reengineered process.

Process reengineering An approach to analyzing and designing processes that calls for a radical overhaul of methods, rather than machines, by first determining the most efficient method of operation and then the technology to implement it.

Process reengineering consultant The individual appointed to the reengineering project team who has specialized knowledge of the process reengineering approach to reforming business processes, including the application of certain tools and techniques.

Real value added (RVA) activity Any business activity or subprocess that accomplishes something that the customer values and is willing to pay for.

Statistical process control (SPC) The application of certain statistical techniques as a means of monitoring the performance of and maintaining control of a particular process.

Structural analysis An approach to analyzing the performance of a process, with an eye toward identifying problem areas, that examines the structural elements of the process rather than its end-to-end performance characteristics.

Systems engineering The process of designing and implementing systems from the perspective of integrating everything that goes into this process, including inputs, outputs, job tasks, work packages, and controls, to produce the desired outcomes.

Team leader The individual appointed as the project manager over the reengineering project.

Value analysis A formal process for isolating the functional attributes of a product or service and then identifying better ways to fulfill these functional requirements by reducing cost and/or improving quality. Also referred to as *value engineering*.

REFERENCES

Blackburn, Joseph D. "Time-Based Competition: White-Collar Activities." *Business Horizons* 35 (July-Aug. 1992), 96–101.

Byrne, John A. "Paradigms for Postmodern Managers." *Business Week,* 1992 Reinventing America Issue, 62.

Diamond, William J. *Practical Experiment Designs for Engineers and Scientists.* Bellmont, Cal.: Wadsworth, 1981.

Dinsmore, Paul C. *Human Factors in Project Management.* New York: AMACOM Books, 1990.

Dorman, George C. "Go with the Flow—Measuring Information Worker Quality." In *Total Quality Performance,* edited by Lawrence Schein and Melissa A. Berman. New York: The Conference Board Research Report no. 909, 1988.

Grant, Eugene L., and Richard S. Leavenworth. *Statistical Quality Control.* 6th ed. New York: McGraw-Hill, 1988.

Guinta, Lawrence R., and Nancy C. Praizler. *The QFD Book: The Team Approach to Solving Problems and Satisfying Customers Through Quality Function Deployment.* New York: AMACOM Books, 1993.

Hakes, Chris, ed. *Total Quality Management: The Key to Business Improvement.* London: Chapman & Hall, 1991.

Hammer, Michael. "Reengineering Work: Don't Automate, Obliterate." *Harvard Business Review* 68 (July-Aug. 1990): 104–112.

Hammer, Michael, and James A. Champy. *Reengineering the Corporation: A Manifesto for Business Revolution.* New York: HarperCollins, 1993.

———. "What Is Reengineering?" *Information Week* (supplement), 5 May 1992, 10–20.

Hammonds, Keith H, and Gail DeGeorge. "Where Did They Go Wrong?" *Business Week,* The Quality Imperative Issue, 25 Oct. 1991, 34–38.

Harrington, H. J. *Business Process Improvement: The Breakthrough Strategy for Total Quality, Productivity, and Competitiveness.* New York: McGraw-Hill, 1991.

Henkoff, Ronald. "Making Your Office More Productive." *Fortune* 123 (25 Feb. 1991), 72–84.

Hequet, Marc. "Quality Concerns Come Through Unfiltered at Johnson Filtration." *The Quality Imperative.* Minneapolis, Minn.: Lakewood Publications, 1992.

Heyer, Steven J., and Reginald Van Lee. "Rewiring the Corporation." *Business Horizons* 35 (May-June 1992), 13–22.

Ignizio, James P. *Goal Programming and Extensions.* Lexington, Mass.: Lexington Books, 1976.

Kelly, Kevin. "A Bean-Counter's Best Friend." *Business Week,* The Quality Imperative Issue, 25 Oct. 1991, 42–43.

Krass, Peter. "A Delicate Balance." *Information Week* (supplement), 5 May 1992, 26–30.

Kreigel, Robert J., and Louis Patler. *If It Ain't Broke . . . Break It!* New York: Warner Books, 1992.

Lacy, James A. *Systems Engineering Management: Achieving Total Quality.* New York: McGraw-Hill, 1992.

Lano, R. J. *A Technique for Software and Systems Design.* Amsterdam: North-Holland Publishing Co., 1979.

Leibs, Scott. "Get Radical." *Information Week* (supplement), 5 May 1992, 6–7.

Meredith, Jack R., and Samuel J. Mantel. *Project Management: A Managerial Approach.* 2d ed. New York: John Wiley & Sons, 1989.

Miller, Jeffery. "Know When to Fold 'Em." *Information Week* (supplement), 5 May 1992, 44.

Payne, Thomas A. *Quantitative Techniques for Management: A Practical Approach.* Reston, Va.: Reston Publishing, 1982.

Roberts, Lon. *Statistical Process Control for Intuitive Thinkers.* Plano, Tex.: Roberts & Roberts, 1992.

Russo, Edward J., and Paul J. H. Schoemaker. *Decision Traps: The Ten Barriers to Brilliant Decision-Making and How to Overcome Them.* New York: Simon & Schuster, 1989.

Schnitt, David L. "Reengineering the Organization Using Information Technology." *Journal of Systems Management* 44 (Jan. 1993): 15.

Steere, Ralph E., Jr. *Office Work Simplification.* Englewood Cliffs, N.J.: Prentice Hall, 1963.

Stewart, Thomas A. "Reengineering: The Hot New Managing Tool." *Fortune* 128 (23 Aug. 1993): 41–48.

Texas Instruments. *Business Process Engineering Concepts.* Plano, Tex.: Texas Instruments, 1992.

Thamhain, H. J., and D. L. Wilemon. "Leadership, Conflict, and Project Management Effectiveness." *Sloan Management Review* 19 (1975): 31–50.

Tylczak, Lynn. *Get Competitive: Cut Costs and Improve Quality.* Blue Ridge Summit, Penn.: Liberty Hall Press, 1990.

Wilder, Clifton. "Measuring the Payoff from Re-engineering." *Computerworld,* 18 Nov. 1991, 65.

INDEX